Thermodynamics Kept Simple

A MOLECULAR APPROACH

Thermodynamics Kept Simple

A MOLECULAR APPROACH

"What is the Driving Force in the World of Molecules?"

Roland Kjellander

CRC Press
Taylor & Francis Group
Boca Raton   London   New York

CRC Press is an imprint of the
Taylor & Francis Group, an **informa** business

CRC Press
Taylor & Francis Group
6000 Broken Sound Parkway NW, Suite 300
Boca Raton, FL 33487-2742

© 2016 by Taylor & Francis Group, LLC
CRC Press is an imprint of Taylor & Francis Group, an Informa business

No claim to original U.S. Government works

Printed on acid-free paper
Version Date: 20150721

International Standard Book Number-13: 978-1-4822-4410-6 (Pack - Book and Ebook)

Visit the Taylor & Francis Web site at
http://www.taylorandfrancis.com

and the CRC Press Web site at
http://www.crcpress.com

Printed and bound by CPI Group (UK) Ltd, Croydon, CR0 4YY

Contents

Preface

The contents of this book is the result of a long process that started at the end of the 1960s when I was an undergraduate student at the Royal Institute of Technology in Stockholm, Sweden, and met thermodynamics for the first time. The excellent teaching in physical chemistry left behind, after all, a troubling void within me: What was thermodynamics really about? The other parts of physical chemistry were focused on molecular properties and processes, while thermodynamics essentially had a macroscopic perspective. A purpose of the use of thermodynamics in chemistry is to be able to make statements and predictions about properties, processes, and equilibria for molecular systems, but where had the molecules gone in the conceptual world of thermodynamics? They were, of course, present in some ways, but they were still pretty peripheral in the treatment of the subject. (This is typical of traditional teaching of thermodynamics and it depends on the nature of classical thermodynamics, which is independent of molecular descriptions.)

I got answers to some of my questions when I was studying statistical thermodynamics in the more advanced courses, but I was still not happy with how the link between microcosm and macrocosm was treated. Most textbooks in statistical thermodynamics have a treatment that is "piggybacking" on the concepts of thermodynamics and its laws instead of giving a concrete explanation of them from a molecular perspective. Surely it should be possible to give a consistently molecular description of thermodynamics in the teaching!

When I started to do research in statistical mechanics, these thoughts about teaching gradually matured and gave rise to ideas on how teaching in molecular thermodynamics could be implemented in practice. During my years in the 1980s as a researcher at the Australian National University in Canberra, Australia, I had the opportunity to test my ideas in practice in a short course for honors students in chemistry. Encouraged by the positive response from the students, I used these ideas to create an advanced course in statistical thermodynamics in 1991 at the University of Gothenburg, Sweden, with successful results.

My feeling was, however, that it should be possible to use the very essence of these ideas to teach thermodynamics at the first-year university level. In 1996, I therefore wrote a small booklet on entropy intended to be used as a supplement to the textbook in our basic course in general chemistry. The response from the students was very positive and we continued to use the booklet in this course. To my delight, it was also used in several other universities and colleges in Sweden. Encouraged by this response I wrote in 2002 a book in Swedish,[1] where I showed how thermodynamics could be introduced from a molecular perspective. With the help of this book, I could fully implement my ideas about the teaching of thermodynamics in our basic course. The students' response was a very satisfying experience for me and it made me feel strongly that I had made a correct choice. Their response led to me being awarded the university's individual Pedagogical Prize 2004 for showing a completely new way of thinking in the teaching of basic thermodynamics.[2] The book was also very well received in the rest of Sweden and this led eventually to the publication of a second edition in 2011.[3] The present book is essentially a translation of this second edition of the Swedish book. Some parts have been reworked and some new contents have been added.

An example of my considerations on thermodynamics teaching is the following insight. The reason why many people have conceptual difficulties with entropy is that the concept of temperature and its properties are taken for granted. Naturally, everyone has an

---

[1] Roland Kjellander, *Vad är drivkraften i molekylernas värld? En molekylär introduktion till termodynamik* ("What is the driving force in the world of the molecules? A molecular introduction to thermodynamics"), Studentlitteratur, Lund, Sweden, 2002.

[2] The university's justification for this prize was, in translation: "Kjellander introduces the concept of entropy early in the teaching, and when seen from a molecular perspective, thermodynamics is perceived as being easy and accessible rather than abstract and elusive, which it does appear in traditional teaching. Kjellander acts as a subject didactics pioneer in this field. He takes the contents of a subject area, starts from the thinking of the students and establishes via his teaching a meeting between students and content. Kjellander emphasizes the students' understanding instead of merely correct answers of the mathematical models."

[3] The title and publisher are the same as for the first edition, see footnote 1. The text of Appendix A in this second edition of the book was with permission from the Swedish Research Council taken from the author's chapter *Vad är temperatur?* ("What is temperature?") in the Research Council's 2003 thematic book *Hett om kalla fakta* ("Hot about cold facts"), Vetenskapsrådet, Stockholm, Sweden, 2003.

everyday experience of temperature, but since this concept is used to introduce entropy (via the second law of thermodynamics), the latter becomes rather mysterious and difficult to comprehend. In fact, entropy is simpler than temperature and is more directly related to molecular properties. In the book I turn it all around. The concept of entropy is introduced first and is connected to probabilities of various molecular processes. Temperature is then introduced and is given an interpretation related to probabilities for heat transfer between different systems. This approach gives the second law of thermodynamics a natural and intuitive background.

An essential part of the philosophy behind the presentation of thermodynamics in the book is to use simple but well-chosen examples to provide insights into the molecular events underlying the thermodynamic macroscopic description. Thereby, insights are also gained regarding the world that we experience ourselves in our everyday life. In the work with the book I have been driven by the following conviction: With the help of our imagination, the world of molecules can also become our own world!

I am deeply grateful to many people during the different phases in the writing of the various versions of the book. Let me first thank the many students and other readers for their constructive and encouraging comments throughout the years. For the Swedish version of the book I want in particular to thank Kjell Johansson, Sture Nordholm, Björn Forsberg, and Kai Lüder for reading the first manuscript and providing constructive comments. Kjell, thank you for your careful review of the text and your positive support. Sture, your unwavering support and encouragement is a great asset for me both professionally and personally. Björn and Kai, your efforts were important to me especially since you at the time had fresh experiences of being students. A huge thanks to Gunnar Numeus for everything you give me when I need it the most and always otherwise too. With your support, it is easy to write a book like this. Last but by no means the least, I want to thank Kim Nygård for carefully reading the English version of the manuscript and for giving constructive input to the text. To have you as a collaborator in scientific research and as a colleague in teaching has given my professional life an extra inspiration.

To the reader

Thermodynamics has an important place in physics, chemistry, and many engineering disciplines. The topic does, however, often give rise to conceptual difficulties partly because of its abstract nature. In particular, the concepts of entropy and free energy have always proved to be difficult to grasp and understand for almost all students. One reason is that these entities are usually introduced in a rather mathematical and abstract manner, which causes this kind of difficulty. Traditional approaches to thermodynamics have these problems almost universally. This book shows that one possible way to provide an understanding of thermodynamics is to give a guided tour of the world of molecules. Thereby, many of the difficulties can be avoided provided the essential principles of thermodynamics are introduced and illustrated from a molecular perspective. Furthermore, it is not necessary to use a lot of mathematics to achieve an intuitive understanding of these matters. An effort has therefore been made in the book to avoid mathematics as much as possible. Instead the focus is on understanding via concrete examples and simple, but appropriate, reasoning. The reader will thereby also acquire an understanding of important phenomena and processes in the macroscopic world.

The book is suitable to use during the first year and upwards for university studies in science subjects. It can be read as a stand-alone book or as a complement to conventional textbooks. The reader should, after having read the book, be equipped with the knowledge and understanding that are needed to fully benefit from more traditional approaches to thermodynamics as dealt with in other literature.

The pedagogical approach and material of the book has been successfully used in first-year chemistry teaching at the University of Gothenburg, Sweden, for many years. The author has also used the material with very successful results in higher level courses for students who have only encountered the traditional treatment of thermodynamics.

Items marked with an asterisk (*) contain optional specialized topics that are not needed for reading the rest of the book. There is a large amount of footnotes that provide in-depth explanations and in many cases more advanced outlooks and comments, including mathematical arguments. At the end of the book there is a list of most symbols used in the book.

To teachers

This book contains a presentation of fundamental concepts and relationships of thermodynamics, including the thermodynamic laws, heat, work, internal energy, enthalpy, entropy, free energy, temperature, pressure, reversible and irreversible processes, the ideal gas law, kinetic theory of gases, heat capacity, standard states, phase transitions, the law of mass action, and the relationship between equilibrium constants and free energy (ΔG^0). In addition, some simple concepts from statistical thermodynamics are introduced and used. The starting points and perspectives are always molecular. Several illustrative and concrete examples are embedded in the presentation.

The book can be used as a complement to existing textbooks or as a stand-alone textbook. Particularly Chapters 2 and 3, which mainly deal with entropy, temperature, and to some extent free energy, are suitable as a complement to the conventional presentation of these concepts given in most textbooks (and that students in general find especially difficult to understand). This part, in particular, is based on pictures and simple reasoning that lead to the correct mathematical relations. The introduction of the concept of entropy is based on a consistently molecular approach and is done before the notion of temperature is treated. This leads to a natural and intuitive background for the second law of thermodynamics. Through numerous concrete examples the reader is prepared for the concept of free energy (Helmholtz energy), which acquires a natural role.

Chapter 4 deals with other parts of the basis of thermodynamics. Traditional teaching in thermodynamics often starts with gas laws (usually as based on empirical findings) and then goes over to energy transfer (heat and work) followed by the concept of enthalpy. Only thereafter entropy and free energy are introduced.[4] The presentation

---

[4]If one wants to pass through the topics in a traditional order one reads about molecular motions, interactions, and internal energy in Section 2.1 followed by work, heat, and the first law of thermodynamics in Sections 4.1 and 4.2, the ideal gas law in Section 4.4 after the shaded box that finishes with Equation (4.13), and heat capacity and enthalpy in Sections 4.5 and 4.6. Then it is time for entropy and the second law of thermodynamics, whereby one continues to read the book from Section 2.2.

in this book takes advantage of the fact that entropy and temperature are already treated when properties of gases are examined, which has several advantages.[5] Thereafter, enthalpy, Gibbs energy and several other quantities and concepts are introduced. The principles of chemical equilibrium in the gas phase are treated in some detail in Chapter 5. In Chapter 6 it is examined what happens when the temperature changes, such as passages of phase transitions (boiling and freezing) and shifts in chemical equilibria. The temperature dependence of entropy is investigated in detail and that of other thermodynamic quantities, treated elsewhere in the book, is summarized. Finally, in Chapter 7 the most important principles discussed in the book are summarized and placed in perspective.

Chapters 4 to 6 contain some reasoning of more mathematical character than in Chapters 2 and 3 and include by necessity some simple derivations, which are clearly marked as such. The structure and focus share, however, the same spirit as the first chapters, and the molecular perspective is identical. Also Chapters 4 to 6 can be used as a supplement to existing textbooks.

The book is written especially for first-year students of science subjects at the university level, but it can be used also at higher levels and as an extracurricular book for particularly interested pre-university students. The presentation is so designed that it gradually gives the reader insight into the conceptual framework of thermodynamics and to some extent its mathematical treatment in the simplest form.

---

[5]For example, the ideal gas law can be *derived* in a quite simple manner that shows that the thermodynamic absolute temperature (associated with entropy) is the same as the temperature in this law.

Author's biography

Roland Kjellander acquired a master's degree in chemical engineering, a Ph.D. in physical chemistry, and the title of docent in physical chemistry from the Royal Institute of Technology, Stockholm, Sweden. He is currently a professor emeritus of physical chemistry in the Department of Chemistry and Molecular Biology at the University of Gothenburg, Sweden. His previous appointments include roles in various academic and research capacities at the University of Gothenburg, Sweden; Australian National University, Canberra; Royal Institute of Technology, Stockholm, Sweden; Massachusetts Institute of Technology, Cambridge, USA; and Harvard Medical School, Boston, USA. He was awarded the 2004 Pedagogical Prize from the University of Gothenburg, Sweden, and the 2007 Norblad-Ekstrand Medal from the Swedish Chemical Society. Professor Kjellander's field of research is statistical mechanics, in particular liquid state theory.

CHAPTER 1

Introduction

What drives chemical reactions forward? What are the conditions for a reaction to be possible? Why is the temperature equalized in an object when a part of it has been heated? What, exactly, *is* temperature? What is heat? Why does a gas become hot when we compress it? What happens when we dissolve a substance in a liquid? Why does fog appear when moist air cools in the evening?

These are some of the issues addressed in this book, which also poses the general question "What is the driving force in the world of molecules?" In many cases there are simple explanations, provided that one is able to use the imagination to understand what happens to the molecules – to "take part" in the molecular world. The idea is to introduce the reader into this way of looking at these issues and to provide intuitive understanding of the individual examples. The principles that are illustrated are, however, universal and can be applied to more complex cases.

The purpose is to explain thermodynamic concepts in a simple, yet correct, way and to place them in a molecular context. Some of these concepts have always proved to be difficult to understand for almost all students. This is well illustrated by the following quote by Arnold Sommerfeld,[1] a famous German physicist:

> Thermodynamics is a funny subject. The first time you go through it, you don't understand it at all. The second time you go through it, you think you understand it, except for one or two small points. The third time you go through it, you know you don't understand it, but by that time you are so used to it, so it doesn't bother you any more.

Thermodynamics is usually presented in a rather mathematical and abstract manner, which gives rise to this kind of difficulty. Statistical thermodynamics, which provides the molecular background and explanation, can be considered as even more mathematically difficult to access. This is probably the most important reason why the

---

[1]Arnold J. W. Sommerfeld (1868–1951) was a German theoretical physicist who made important scientific discoveries in atomic and quantum physics.

molecular background of thermodynamics is often avoided in basic teaching. This book shows that a lot of mathematics is not always required to obtain an intuitive, molecular understanding of the essential principles of thermodynamics. Instead, the presentation is largely based on pictures and fairly simple reasoning that leads to the correct mathematical relationships. This makes the book unique in many respects. Ordinary textbooks treat thermodynamics in a way that is unnecessarily abstract and complicated, at least for students who for the first time are studying the subject. This is unfortunate because the molecular events that underlie thermodynamics are fascinating and, in addition, they are vital for modern chemistry and physics. Thermodynamic principles govern so much of what happens in nature, so an intuitive understanding of the molecular background is very desirable for all science students and professionals.

We will introduce and explain thermodynamic principles by using a number of concrete examples. For instance, we shall treat:

- Expansion and compression of gases

- Mixing of gases with each other

- The spreading of energy in a body

- Equilibrium between a solution and the solid phase for sparingly soluble substances

- Evaporation of a liquid droplet and the appearance of equilibrium with saturated vapor

- Some chemical reactions that take place under release or absorption of energy

- Appearance of equilibrium for chemical reactions

- Interactions between charged bodies in an electrolyte solution

- The principles of a refrigerator from a molecular perspective

- The molecular basis of pressure of gases and other gas properties

- Melting of solid phases and freezing of liquids

- Boiling of liquids and condensation of vapor

A major theme is to examine what is the driving force in microcosm, that is, what is really going on molecularly that makes spontaneous processes go forward. We ask ourselves, for example, the following questions:

- What makes two gases spontaneously mix with each other when they are brought into the same container and why don't they separate again spontaneously?

- Why does energy in an object spread after heating, so that the object becomes equally warm everywhere after a while?

- Why is at least a small amount of a substance always soluble in another substance? What has happened when equilibration has taken place between a solution and the solid phase of a sparingly soluble substance?

- Why does a drop of liquid evaporate until the vapor becomes saturated and what has happened then?

- What drives the process forward when we burn a candle and when magnesium burns in fireworks?

- How is it possible that there are spontaneous chemical reactions that take up energy, i.e., it becomes cold when they occur?

We will find that there is a common denominator in all these different examples, a quantity the discovery of which was one of the great triumphs of thermodynamics. This quantity is called **entropy** and its behavior determines whether a process can occur spontaneously or not. Entropy is actually a rather abstract entity, which can appear puzzling, but we shall see that its existence and properties are very natural. A prerequisite to see this is that we mentally "visualize" the course of events that the molecules are involved in; that we with the help of our imagination and thoughts "participate" together with them in a world of which we do not have an everyday experience. So ... Welcome to the world of molecules!

Goal

- An understanding of spontaneity, entropy, free energy, temperature, and other thermodynamic concepts from a molecular perspective.

CHAPTER 2

Energy and entropy

2.1   In the world of molecules
Movements, interactions, and energy

Most of what we see and feel in our world, the macroscopic world, is the result of what happens in the microscopic world – the world of molecules. The molecules are so small that we cannot observe them directly with our senses and what we perceive depends on the course of events with many molecules simultaneously involved. For example, when the wind blows, the force that one experiences on the body originates from molecules that collide against the body surface.[1] During a gale, the force may be so great that one cannot stand still. When there is no wind, equal forces act on either sides of the body due to the air pressure,[2] so the net force is zero and one is not aware of it. The force on each side is, however, very large. At atmospheric pressure, molecular collisions give rise to a force of $10 \, \mathrm{N \, cm^{-2}}$, meaning that on each square centimeter there acts a force that corresponds to the weight of about 1 kg. On the whole body surface of an adult, the force corresponds to a weight of more than ten thousand kilograms! Because the forces are so large, just a relatively small difference in the collision intensity on either sides of the body is needed in order to give a net force that is considerable, as in a gale. But why are we not strongly compressed by the gas pressure when the forces on the body are so great? The reason is that our tissues have an internal pressure which precisely counteracts the external pressure.

Molecular collisions thus give rise to the gas pressure acting on a surface. Every single collision gives a very small contribution to the force on the surface, but because there is an enormous number

---

[1] A collision of a molecule with the body surface gives rise to a force that acts on the body. It is like being hit by a ball, but on a much smaller scale. The force from *each* molecular collision is tiny since molecules have very small masses.

[2] The pressure on the body surface is equal to the force per unit area due to collisions by air molecules. The concept of pressure is treated in more detail in Section 4.1.

of molecules that collide with the surface at about the same time,[3] the total force that the gas molecules inflict on the surface is very large. The very fact that the molecules are so numerous also means that the total force in practice does not vary noticeably over time, but is constant at equilibrium – one cannot distinguish individual molecular collisions and what is perceived is the average of the total force.

From a macroscopic perspective one does not perceive the movements of the molecules in a gas at equilibrium. A nitrogen molecule moves at room temperature with an average speed[4] of about 500 meters per second. However, it does not move 500 m away from the starting point during a second because it collides with other gas molecules on the way. It moves in a tortuous zigzag path; at atmospheric pressure, it moves on average less than 0.1 micrometers in a straight line (a few hundreds of molecular diameters) before it collides with another molecule and changes direction. For similar reasons, we can understand that when someone opens a bottle with a fragrant substance, it takes time for the molecules to spread a few meters so one can smell it from a distance. The time it takes depends on how the paths of the scent molecules are affected by the molecules in the air. Furthermore, the speed of the molecules depends on their mass. A heavy molecule moves on average slower than a lighter one at a given temperature.

That a molecule moves forward at a certain speed (so-called translational motion) is only *one* kind of motion it might have. In addition it rotates and vibrates, so-called rotational and vibrational motions. On average, the molecules have faster motions when the temperature is increased; that is, they move forward faster and rotate and vibrate more violently. These movements are often called thermal motions. The faster the motion, the higher the kinetic energy (the energy of motion).

When two molecules are close together they affect each other by forces that may be attractive or repulsive. In other words, the molecules interact with each other. These forces may be due to electrical charges on the molecules, so-called electrostatic interactions, or be due to some other reasons. The fact that molecules have size, i.e., they repel each other strongly at contact, is also due to a kind of

---

[3]At each square centimeter there are about $3 \cdot 10^{23}$ collisions per second (for air at 25°C and atmospheric pressure).

[4]In this book we distinguish between the concepts of speed and velocity. The velocity has a direction while speed is the magnitude of the velocity (the length of the velocity vector).

interaction. In a liquid, where the molecules are close together, each molecule interacts with many molecules simultaneously, while in a gas at atmospheric pressure one can ignore the interactions for each molecule, except at the moment it collides with another molecule.

The molecules collide and exchange energy with each other by interactions all the time. Thereby the motion of each molecule is changing again and again in rapid succession. Normally both the speed and direction of motion changes for the colliding molecules. The speed of one molecule is thereby reduced while the speed of the other is increased. The *total* energy of the molecules is, however, the same before, during, and after the energy exchange. This is in accordance with an important law of nature that says that the sum of all the energy is constant. This law is called the **first law of thermodynamics**, and it means that energy can neither be destroyed nor created, but only converted between different forms of energy. Apart from kinetic energy, the molecules also have potential energy. The latter form of energy has to do with the interactions between the molecules and between the particles that constitute them (electrons and nuclei). If, for example, the distance is increased between two particles that attract each other, their potential energy will increase, just as gravity makes the potential energy of a weight increase when one lifts the weight from the surface of Earth. Likewise, the energy is increased when the distance is decreased between two repelling particles.

Energy is transformed continuously between the various forms since molecules move and interact with each other.[5] For example, if the kinetic energy of the molecules is changed, their interaction energy will also be changed on average; moreover the energy of electrons in the molecules changes in general. Therefore, both the kinetic and potential energies of matter increase on average when one adds energy. Hence it is not only thermal motions that increase with increased temperature but also, for example, the energy "inside" the molecules.

At each instant of time the molecules have different speeds and different energies. For example, the nitrogen molecules in air have different speeds: some move slowly, many have a higher speed, and relatively few move very quickly. Most molecules have a speed fairly

---

[5]If, for example, two molecules move towards each other, an attraction between them will cause an acceleration towards each other and, correspondingly, a repulsion will cause a reduction in speed. The strength of the attraction/repulsion is simultaneously changed because the interaction depends on the distance between the molecules.

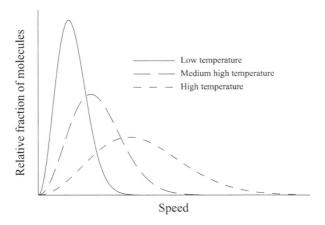

Figure 2.1   The distribution of molecular speeds at various temperatures for molecules of the same mass. The curves show the proportion of molecules that have a certain speed.

close to the average speed, which depends on the temperature and on the mass of the molecules. Some typical distributions of speed of the molecules at various temperatures are shown in Figure 2.1.[6] We see that at high temperatures there are many molecules that move much faster than most molecules do at lower temperatures. Yet, at all temperatures there are always some molecules that move slowly and few that move very quickly, even if the number in the last category is vanishingly small at low temperatures. When the temperature is increased the distribution becomes wider, that is, the difference becomes greater between the low and high speeds that occur for many molecules.

This distribution of molecular speed implies, of course, that the kinetic energy of translational motion for the molecules is distributed in a similar manner (this energy is proportional to the square of the speed).[7] Other forms of energy are also distributed among the molecules so that some have low energy, many have higher energy, and relatively few have a really high energy. The higher the temperature, the greater the fraction of molecules with high energy, but there

---

[6]This distribution of speeds of molecules is described in more detail in Appendix B, which contains optional specialized material.

[7]The translational energy is $\varepsilon_{tr} = mv^2/2$, where m is the mass and v the speed of the molecule.

are always some with relatively low energy. Most have an energy fairly close to the average energy per molecule.

The fact that there are always a few molecules that have higher energy than most of the others has important consequences. For example, this makes it possible to dry laundry at low temperatures and makes water dry up on the ground after rain. High temperature is, in fact, not needed for molecules in a liquid to be released from the surface of a liquid and hence vaporize. A liquid is "held together" by attractive forces between the molecules and energy is therefore required to release a molecule from the attraction to the others. Since some molecules have higher energy than others, there are always a few that have enough energy to break away from the attraction and turn into gas. If the humidity is not too high, the evaporation will continue in this manner and the laundry and street will eventually become dry. The energy required is taken mainly from the surroundings that will be somewhat colder. One can even dry laundry at freezing conditions when the water in the wet laundry is frozen. Even in ice there are some molecules that have sufficient energy to vaporize. One says that the ice sublimates when it is vaporized. However, it will take considerable time because there are very few molecules that vaporize at any instant in time, but the laundry nevertheless becomes dry in the end.

Other examples of the importance of molecules with higher energy than the majority can be found among most chemical reactions. There is usually an energy barrier that separates the reactants (the starting substances) and the products; that is, there is an intermediate stage that must be crossed during the reaction and that has higher energy than the reactants and products. When a collision takes place between an energetic molecule and another molecule, the former can supply the energy needed for the barrier to be crossed and a reaction to occur. A reaction can thus occur even if most molecules have too low energies to be able to pass the barrier. Sometimes the barrier is, however, so high that it is very unlikely that there are molecules with high enough energy to pass it. Then one may need to bring energy from "outside" to give some molecules sufficient energy, such as when one ignites the gas from a Bunsen burner with a match. Once the gas is burning, the energy needed to ignite more gas is supplied by the gas flame that already exists and no energy is needed to be supplied from outside.

Before concluding this section, we will introduce some useful definitions. Henceforth, we will use the concept of a **system**. A system is simply the part of the universe that we are interested in; for example a gas enclosed in a box, a reaction mixture in a flask, a biological cell, or a bay of a lake.[8] The total energy of a system is denoted U and it is the sum of all kinetic energy (translational, vibrational, and rotational energy) and potential energy (interactional energy) of the molecules in the system. Since U is the energy content of the system itself, it is known as the **internal energy**. The internal energy can be changed by transferring energy *to* the system, whereby U increases, or *from* the system, whereby U decreases. When energy is transferred between two systems, one system gains equal energy as the other one loses or vice versa, since energy can neither be destroyed nor created. For an **isolated system**, energy and particles *cannot* pass the system boundary. Therefore the internal energy of the system is constant, and if no chemical reactions occur, the numbers of particles of various species are also constant. For a **closed system** no particles can pass the boundary, but energy can.

Matter exists in various states of aggregation: the most important are the gas, liquid, and solid states. One speaks about various phases of a substance; a **phase** is a form of matter that is uniform throughout space in its physical state and chemical composition. In a solid phase the molecules are immobile relative to each other – in a crystalline solid the atoms/molecules are sitting on definite places in space relative to each other in a pattern that is repeated in a periodic manner (a crystal lattice). The atoms/molecules can, however, vibrate while sitting on these places. In a liquid phase the molecules are free to move relative to each other, but they are held densely together by attractive intermolecular interactions. Both a liquid and a solid assume a certain finite volume. For a gas phase the molecules are not held together so the gas fills the entire space that is available for it. For a thin gas the interactions between the molecules are negligible except when they collide with each other. This is the case for a gas at normal pressures and temperatures (about one atmosphere and room temperature).

Key points

- The internal energy U of a system is the sum of potential and kinetic energy of the atoms and molecules in the system.

---

[8]The macroscopic systems we consider in this book are always at rest as a whole; only parts of each system may move.

- The energy of individual molecules is continuously transformed among the different forms of kinetic and potential energies when the molecules move, collide, and interact with each other.

- At any instant of time the molecules have different energies. Some have low energy, many have higher energy, and relatively few have very high energy. Most have an energy fairly close to the average value per molecule of each species.

- Energy can neither be destroyed nor created. When energy is transferred between two systems, one system gains equally as much energy as the other one loses.

2.2 Self-evident matters?
Spreading and spontaneity

Take a box with a removable partition that separates a smaller part of the volume from the rest. Let us enclose a gas inside this part, while the rest of the box is empty as in Figure 2.2.

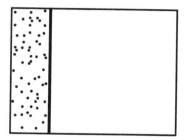

Figure 2.2   A box with a partition that prevents the particles in the left part from reaching the right part, which is empty.

The gas molecules, which we for simplicity will call "particles," move freely within the smaller volume. Each particle is moving straight forward until it collides with something – a wall or another particle – which causes it to change direction, after which it again moves in a straight line. All particles remain in the smaller volume, since the collisions with the partition prevent them from reaching the other part of the box.

If we now remove the partition, the particles will spread out in the entire volume. Particles that move to the right are no longer hindered

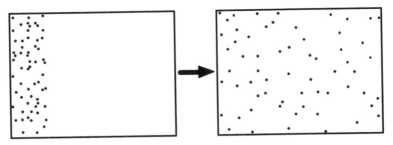

Figure 2.3   All particles are located in the left part of the box immediately after the removal of the partition, but then they spread evenly throughout the whole volume.

by the partition, so they will move into the other part of the volume. After a while, the particles will be evenly distributed throughout the entire volume (see Figure 2.3).

This is an example of a **spontaneous process**. To start with, just after the removal of the partition, all particles are in the smaller volume. Without influence from us or anything else outside of the box, they will spontaneously spread out in the entire volume. The initial state is a gas with a high density in the left part of the box (and density zero in the right part) and the final state is a gas with a lower but uniform density throughout the entire box.

We would be quite surprised if the opposite were to happen: if the gas particles initially were evenly distributed throughout the volume and all particles spontaneously gathered in the left part. Intuitively, it is pretty obvious that such a course of events is very unlikely – although in principle it is possible that it could happen. The more particles we have in the box, the more unlikely it is that they would spontaneously gather in the left part. If we had only two particles in the box, we realize easily that it would actually be quite probable that both would be in the left part at the same time. For three particles, it would be less likely that all three would be so and when we increase the particle number, it quickly becomes very unlikely that all of them simultaneously would be there. We are normally interested in systems with many particles and for such systems the probability is *vanishingly small* that it would happen – in fact, it will never happen.

The reasoning is typical for a spontaneous process. The probability that the process goes in one direction is much greater than that it

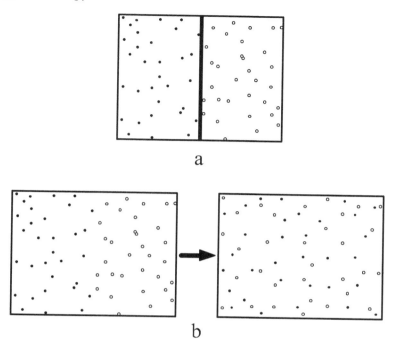

Figure 2.4 (a) A box with one kind of particles at one side of a partition and particles of another kind at the other side. The partition prevents the particles from reaching the opposite part of the box. (b) The two kinds of particles are found in the respective part of the box immediately after the partition has been removed, but then they spread evenly throughout the volume.

would go in the other direction. What are the odds for the process to go one way or the other? This we shall see soon.

A similar example is when a partition initially delimits two different kinds of particles, black on the left and white on the right. Particles of either kind move freely within each volume but are hampered by the partition from reaching the other part of the volume as shown in Figure 2.4a. If we remove the partition, white particles will move into the left part and black into the right part. After a while, both kinds of particles are distributed evenly throughout the entire volume (Figure 2.4b).

The end result is therefore that the two kinds of particles are evenly mixed. In this case, the initial state is that the two gases are separated and the final state is that they are completely mixed. The process is spontaneous. The probability for the reverse process to happen – that the white particles would spontaneously gather in the right and the black of the left half – is vanishingly small. Note that the mechanism behind the process is the same as in the previous example: each gas fills the available volume, since nothing prevents it from doing so.

2.3   Particle locations
Macroscopic and microscopic states

How then shall we be able to quantify the above? How do we express that it is more likely that all particles are uniformly distributed throughout the volume rather than located within a limited area? Let us take note of the places where the particles are at a given moment. In the example with one kind of particles it may be as in Figure 2.5.

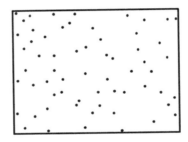

Figure 2.5   A snapshot picture of the particles in a box. We can observe on which places the various particles are, i.e., what configuration they have.

We say that the particles have a particular **configuration**, i.e., that they among themselves and relative to the walls are at certain locations. If we move one or more particles, we obtain a different configuration. When time passes, the particles move and their configuration changes; they go from one configuration to the next and then to the next et cetera.

Let us for simplicity assume that it is an **ideal gas**, that is, a gas in which the particles do not interact with each other. This means that they do not feel each other's presence no matter how close together they are. They thereby behave as if they do not have any size, so-called point particles. In a real gas the particles collide occasionally with each other, but if the gas density is low enough, one can ignore

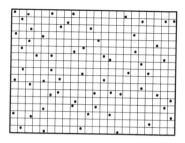

Figure 2.6 A particle configuration is determined by taking note of in which cells the particles are located. The position of the respective particle within each cell does not matter.

this and as an approximation treat the gas as ideal. Thus we ignore collisions and other interactions between the particles.[9]

We will now consider the various particle configurations that can occur. If we have a configuration and a particle is moved a very small distance, say an arbitrarily small fraction of an Ångström,[10] should we count this as a new configuration? In that case, the number of possible configurations would be infinitely many. The answer is, fortunately, no. Quantum mechanics sets limits on whether it is meaningful to distinguish between two different positions of a particle. At a given temperature, the particles have a certain average speed, and the Heisenberg uncertainty principle[11] says that one cannot simultaneously determine the position and the velocity of a particle with arbitrarily high accuracy. This means that it is not meaningful to distinguish between particle positions that are very close to one another.

---

[9]In an ideal gas it is nevertheless assumed that the molecules can exchange energy with each other, at least if sufficient time elapses. This really means that the molecules, after all, have to collide now and then, but for an ideal gas one disregards any other effect of the collisions.

[10]1 Ångström (Å) $= 10^{-10}$ m. This length unit was introduced by the Swedish physicist Anders Jonas Ångström (1814–1874).

[11]The Heisenberg uncertainty principle is a result of quantum mechanics. It implies that if the velocity of a particle in, say, the x direction is determined with an accuracy Δv_x and its location simultaneously determined with a precision Δx, then the following holds: $\Delta v_x \times \Delta x \geq \alpha > 0$, where α is a number of the order h/m, where h is Planck's constant and m is the mass of the particle. Both Δx and Δv_x can hence not be zero simultaneously, and if one is zero the other is infinitely large (the latter quantity is then completely indeterminate). This means that one cannot exactly determine both the position and the velocity of a particle.

Let us therefore divide our volume in small imagined cells as in Figure 2.6 and count all positions of a given particle inside a cell as equivalent, that is, if we move a particle within a cell, this does not count as a new configuration.

A closer analysis shows how large the cells should be,[12] but the precise value of the cell size is rather unimportant. As we shall see, the main results are independent of this value. Note that the cells are not real – the particles are unaware of the cell boundaries. Their purpose is solely to help us to decide which configuration the particles have at every moment. In Figure 2.7, we see some examples of different configurations for our system.

An important fact is that every possible particle configuration for an ideal gas is *equally probable*. This means that if we have 1000 possible configurations, the probability of observing each of them is 1/1000. By determining the number of configurations, we can therefore determine the probabilities that we need in the discussion about distribution of particles that we began earlier.

How many configurations are there? Well, basically one just has to draw them and then count how many there are. We demonstrate this with a simple example with only six cells and start with a system with only *one particle* (see Figure 2.8). The number of configurations is denoted by Ω.

There are six different ways to place a particle in six cells and thus there are six configurations of the system, $\Omega = 6$. For *two particles* there will be more configurations. If one particle is in the first cell, the other particle can be in any of the six cells (see first row in Figure 2.9). The same applies if the first particle is in the second cell (see second row of Figure 2.9), and so on. We allow two particles to be in the same cell because they are point particles. Overall, we have $6 \times 6 = 36$ possible configurations, that is, $\Omega = 6^2$.

For *three particles* we have the following: If one particle is in the first cell, the other two can assume 6^2 configurations, i.e., the configurations we obtained for two particles (see Figure 2.10). The same applies if the first particle is in the second cell, and so on. Overall, we have $6 \times 6^2 = 216$ possible configurations, $\Omega = 6^3$.

We see that the number of configurations grows rapidly with the number of particles, and it becomes impractical to draw them all.

---

[12]The cell size depends only on the particle mass and the temperature. At room temperature, each side of a cell is a fraction of an Ångström.

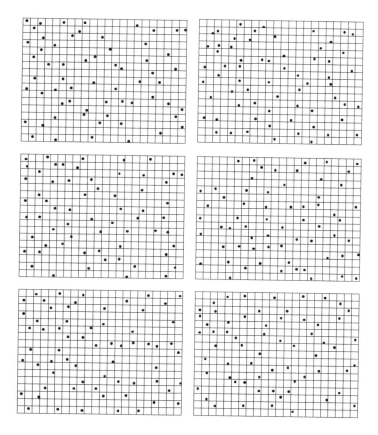

Figure 2.7   Some possible particle configurations. Each of these is equally likely for an ideal gas.

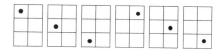

Figure 2.8   A system with one particle and six cells has six possible configurations, $\Omega = 6$.

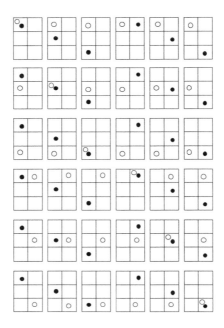

Figure 2.9   A system with two particles and six cells has 36 possible configurations, $\Omega = 6^2$.

However, we can already see the pattern. Each new particle provides a factor of 6, so the number of configurations for N particles is 6^N.

If each cell has a volume v (determined by the particle mass and the temperature), the total volume $V = 6v$ in this case. In the general case for a system with volume V, the number of cells is equal to V/v. If we have one particle there are therefore V/v configurations, for two particles $(V/v)^2$, and for N particles we have

$$\Omega = \left[\frac{V}{v} \right]^N = \frac{V^N}{v^N}. \tag{2.1}$$

This means that if we are interested in how the number of configurations depends on the volume of the system, we can use the formula

$$\Omega = \mathcal{K} V^N, \tag{2.2}$$

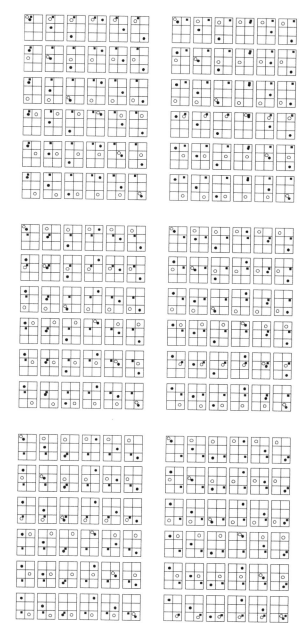

Figure 2.10   A system with three particles and six cells has 216 possible configurations, $\Omega = 6^3$.

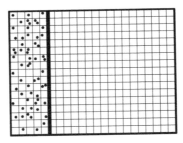

Figure 2.11   Sixty particles are trapped in the left part of a box. The partition prevents them from entering the right part, which is empty.

where $\mathcal{K} = 1/v^N$ is a constant[13] that is independent of the volume V. A doubling of the volume would therefore lead to Ω increasing from $\mathcal{K}V^N$ to $\mathcal{K}(2V)^N$. Since $\mathcal{K}(2V)^N = 2^N \times \mathcal{K}V^N$, we see that Ω has increased by a factor of 2^N.

In the same manner, we see that when we change the volume from V_{before} to V_{after} the number of configurations Ω will increase from $\mathcal{K}(V_{\text{before}})^N$ to $\mathcal{K}(V_{\text{after}})^N$, and therefore we have

$$\frac{\Omega_{\text{after}}}{\Omega_{\text{before}}} = \frac{\mathcal{K}(V_{\text{after}})^N}{\mathcal{K}(V_{\text{before}})^N},$$

which we can write as

$$\Omega_{\text{after}} = \left[\frac{V_{\text{after}}}{V_{\text{before}}}\right]^N \Omega_{\text{before}}. \tag{2.3}$$

Thus, Ω increases by a factor $[V_{\text{after}}/V_{\text{before}}]^N$. Note that this factor is independent of the volume v of the cells. The conclusion is that when the volume increases by a factor $\mathcal{F} = V_{\text{after}}/V_{\text{before}}$, then Ω increases by a factor $\mathcal{F}^N$.

Now, we have learned enough to be able to make a calculation for our first example. Let us examine the box with a partition that has

---

[13]This result applies if the particles are distinguishable, for example if one is black, one is white, and so on. If the particles are indistinguishable from each other (which applies to molecules of the same kind) the constant $\mathcal{K}$ becomes slightly different, but it is still independent of the volume provided V is so large that the number of cells greatly exceeds the number of particles. Apart from the value of $\mathcal{K}$ the formula is still valid. It is also valid as an excellent approximation for particles with size (i.e., that are not point particles as assumed earlier) provided that the volume V is so large that the probability is small for two particles to be near each other.

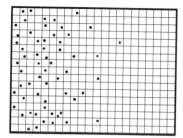 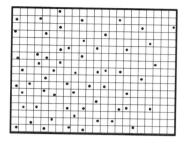

Figure 2.12   The particles spread throughout the volume when the partition is removed, i.e., the gas expands spontaneously.

particles only on one side as shown in Figure 2.2. We first divide the system into cells (see Figure 2.11). When we remove the partition, we increase the available volume for the particles by a factor of 4.4 in this example. Since we have 60 particles, the number of possible particle configurations for the system will increase by a factor

$$4.4^N = 4.4^{60} \approx 4 \cdot 10^{38}$$
$$= 40000000000000000000000000000000000000 \, (!).$$

Already for as few particles as this, Ω increases by a very large number.[14] If we had 1 mole of particles in our system and the volume increased by a factor of 4.4, Ω would increase by a factor of $(4.4)^{6 \cdot 10^{23}} \approx 10^{4 \cdot 10^{23}}$, that is, one followed by $4 \cdot 10^{23}$ zeros, which is an overwhelmingly huge number.

Thus, the system received a very large number of new possible configurations when we increased the volume. The initial configurations, that is, when all particles are located in the left portion of the volume, are still available, and each of them is as likely as any other configuration. However, they are very few compared to all the others, so the chance that the particles would return to any of the original configurations in our example is 1 in $4 \cdot 10^{38}$ – in other words extremely small. For one mole of particles the probability is in practice absolutely nonexistent.

---

[14]This number is about a billion times larger than the estimated age of the universe measured in picoseconds (10^{-12} s). During a picosecond two small molecules like N_2 would just have had time to pass each other in their motions at room temperature. Hence, an extremely tiny fraction of all new possibilities after the volume change will have had time to be explored during this age.

If we watched the system after we had removed the partition, we would see that the particles eventually become spread throughout the volume, and we would thus see a sequence of particle configurations that more and more fill the volume uniformly as in Figure 2.12. The reason for this is that there are enormously many more configurations where the particles are spread out evenly, than there are configurations where many of them are gathered in a small region. It is therefore very likely that the particles are spread evenly throughout the volume. When this has occurred, deviations from uniform distribution are unlikely.

For a macroscopic system, the number of configurations with particles spread evenly is *tremendously* larger than those with particles unevenly distributed. When the particles fill the entire volume, even small deviations from a uniform distribution are *very* unlikely.

What we perceive as a spontaneous process – that the particles at the beginning are gathered in one part and then spread out into the entire volume – is thus a natural consequence of the fact that it is vastly more likely that the particles are evenly distributed. We go from an initial state with relatively few configurations Ω_{before} (before we remove the partition) to a state with many more configurations Ω_{after}. The spontaneous process and the increase of Ω accordingly accompany each other. When Ω for the final state is larger than for the initial state, the process goes spontaneously in this direction. We can express this as the statement: Ω *increases for a spontaneous process.*[15] This is a very important observation to which we shall return.

Let us consider the same process from the perspective of an individual particle. Since the particle does not interact with the other particles, it is completely "unaware" of their existence. It behaves as if it were *all alone in the box* and moves forward in a straight line until it collides with a wall, where it changes direction.[16] For the particle, it does not matter at all if it is in a region with many particles or just a few. It just moves forward, "following its nose," until it hits a wall, completely unconcerned about the environment in general. Thus, it is not so that the individual particles try to avoid being in areas with

---

[15]Thermodynamics in its classic form is concerned only with equilibrium states and not really with the transition processes between them. Both the initial and final states that we are dealing with here are equilibrium states and the increase of Ω for a spontaneous process in this context means that Ω for the macroscopic state after the process is greater than for the initial macroscopic state.

[16]In a real gas, the particles also collide with each other (see also footnote 9).

many particles, but the end result is still that they spread out evenly in the whole volume for reasons of probability. The spontaneous process where particles spread and Ω is increased, is thus *a property of the system as a whole* and not of the individual particles, i.e., an example of how the whole can be more than the sum of its parts.

The end result – that the particle density (the number of particles per unit volume) is the same throughout the volume – is the *macroscopic equilibrium state* of the system. At equilibrium the macroscopic properties of the system do not change anymore. The individual particles of course continue to move, so the particle configuration of the system changes all the time. Although the system goes from one microscopic state (configuration) to another and then to another and so on in an uninterrupted sequence, it seems to us as if nothing changes macroscopically. This is because the overwhelming majority of these microscopic states are virtually identical from a macroscopic perspective, they are macroscopically indistinguishable. At the macroscopic level, we do not see the individual particles, but we perceive only the *average* of the microscopic quantities. For example, the particle density is macroscopically constant throughout the box, while at the microscopic level, at every moment of time the density is zero everywhere except where a particle is currently located – a highly discontinuous function.

A macroscopic (thermodynamic) equilibrium thus corresponds to a huge number of microscopic states ("microstates"), and most thermodynamic properties (such as particle density, number of particles, energy, pressure) are averages of microscopic quantities. This applies in general.

Key points

- Each macroscopic (thermodynamic) state corresponds to a huge number of microstates (in this case the particle configurations) that the system is continually switching between.

- The equilibrium state is the most probable macroscopic state and corresponds to by far the largest number of microstates.

- A spontaneous process from one macroscopic state to another occurs when the number of available microstates, Ω, of the latter state is greater than of the former, that is, Ω *increases for a spontaneous process.*

- When the volume of an ideal gas is increased from V_{before} to V_{after}, the number of particle configurations increases by a factor $[V_{after}/V_{before}]^N$, where N is the number of particles.

2.4   Two independent systems
The concept of entropy

We have seen that the number of particle configurations plays a very significant role when we consider the expansion of a gas. The number of configurations is the number of different arrangements of particle positions within a given volume, that is, the number of different possibilities that the system has access to at the microscopic level. Generally, we let Ω denote the number of different microscopic alternatives (microstates) that are available for the system under the prevailing conditions. Various particle configurations are just one kind of possible alternatives (more to come later).

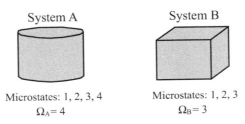

Figure 2.13   Two systems A and B.

Let us consider two systems A and B. Say that system A has four microstates available ($\Omega_A = 4$) and B has three such states available ($\Omega_B = 3$) (see Figure 2.13). Let us now consider the two systems A and B together as a system AB; we thus combine the systems into a single system without modifying either A or B (see Figure 2.14).

How many microstates does the combined system AB have? If system A is in state 1, system B can be in either state 1, 2, or 3 (see first line in Figure 2.15). Similarly, if system A is in state 2, system B can be in state 1, 2, or 3, and so on. In total, therefore, the system AB has $4 \times 3 = 12$ available states:

$$\Omega_{AB} = \Omega_A \times \Omega_B. \qquad (2.4)$$

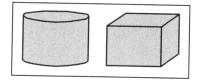

Figure 2.14   The combined system AB.

Figure 2.15   The combined system AB has $\Omega_{AB} = 4 \times 3$ states.

We can easily realize that this relationship applies generally: the number of microstates of a combined system is the product of the number of microstates of the constituent subsystems.[17]

The energy of the combined system AB is, of course, the sum[18] of the energy for A and for B:

$$U_{AB} = U_A + U_B \tag{2.5}$$

Here we see a difference. The energy U is additive, while Ω is multiplicative. As we have seen, Ω plays an important role for instance during a spontaneous process, and it is, in fact, quite impractical to

---

[17]This is strictly true provided the subsystems are unchanged when being combined.

[18]This is strictly true provided the total energy is unchanged when the systems are combined, such as in this case when each subsystem is unchanged.

have a quantity that is multiplicative. But there is a simple solution to this. If we follow Boltzmann[19] and define **entropy** S as[20]

$$S = k_B \ln \Omega \tag{2.6}$$

where k_B is a constant, we obtain a quantity that is additive

$$
\begin{aligned}
S_{AB} &= k_B \ln \Omega_{AB} = k_B \ln[\Omega_A \Omega_B] \\
&= k_B \ln \Omega_A + k_B \ln \Omega_B = S_A + S_B,
\end{aligned}
$$

that is,

$$S_{AB} = S_A + S_B. \tag{2.7}$$

Thereby, the entropy of the combined system AB is the sum of the entropy of A and B. Note that this does not really change anything of substance. If we know how Ω behaves, we also know how $\ln \Omega$ behaves. If Ω increases during a spontaneous process, $\ln \Omega$ increases too. And the same is true if we multiply $\ln \Omega$ with a positive constant k_B: the entropy $k_B \ln \Omega$ also increases. It is simply more convenient to work with S than with Ω.

The constant k_B is called **Boltzmann's constant** and, for historical reasons, its value is $1.38 \cdot 10^{-23}$ J K^{-1} (where J stands for Joule and K for Kelvin). This is related to what temperature scale one wants to use, and because in science one has chosen the Kelvin scale, which in turn is related to the Celsius (centigrade) scale,[21,22] one can show that the constant has this value. The unit of entropy is the same as the unit for k_B since $\ln \Omega$ is a dimensionless number.

---

[19]Ludwig Boltzmann, an Austrian scientist who lived 1844–1906, made major scientific contributions concerning, among other things, the molecular interpretation of thermodynamics. He found the relationship between the thermodynamic concept of entropy and the number of microstates, $S = k_B \ln \Omega$.

[20]The concept of entropy was introduced in classical thermodynamics much earlier than Boltzmann's expression for it. For all practical purposes, Boltzmann's entropy for a macroscopic system agrees with the classical entropy. In Section 4.3 we will discuss some aspects of entropy in classical thermodynamics.

[21]The temperature scales are related such that if Θ is the temperature in °C and T is the temperature in Kelvin (K), $T/K = \Theta/°C + 273.15$.

[22]The Celsius scale was introduced by Anders Celsius (1701–1744), a Swedish astronomer who also worked with geographical and meteorological measurements. Originally his thermometer scale had 0 at the boiling point for water and 100 at the freezing point. The scale was later reversed to its present form.

We saw earlier that Ω for an ideal gas depends on the volume V according to Equation (2.3) when we change the volume from V_{before} to V_{after}. From this we can obtain an expression for the change in entropy during a volume change. If we take the logarithm of both sides of Equation (2.3) we obtain

$$\ln\Omega_{\text{after}} = \ln\left[\frac{V_{\text{after}}}{V_{\text{before}}}\right]^N + \ln\Omega_{\text{before}} = N\ln\frac{V_{\text{after}}}{V_{\text{before}}} + \ln\Omega_{\text{before}},$$

where we have used the logarithm rule $\ln a^b = b\ln a$. Since the entropy change ΔS is given by

$$\Delta S = S_{\text{after}} - S_{\text{before}} = k_B\ln\Omega_{\text{after}} - k_B\ln\Omega_{\text{before}}$$

we finally obtain the result[23]

$$\Delta S = Nk_B\ln\frac{V_{\text{after}}}{V_{\text{before}}} = nR\ln\frac{V_{\text{after}}}{V_{\text{before}}}, \tag{2.8}$$

where $n = N/N_{\text{Av}}$ is the number of moles, N_{Av} is the Avogadro constant and $R = N_{\text{Av}}k_B$ is called the **universal gas constant**.[24] Notice how much nicer it is to have a factor of N in the expression for ΔS than having the exponent N in the expression (2.3) for Ω. If we double the number of particles $\Delta S = S_{\text{after}} - S_{\text{before}}$ will be doubled, while $\Omega_{\text{after}}/\Omega_{\text{before}}$ would be raised to the second power. Certainly it is more convenient to deal with additive functions than with multiplicative ones!

The relation (2.8) between entropy and volume changes of an ideal gas is an example of a relationship that has major implications for chemistry and physics. Similar relationships also exist for other kinds of systems. Equation (2.8) gives the change in **configurational entropy**, S_{conf}, for an ideal gas, that is, the entropy that depends on the number of configurations available for the particles. We therefore write

$$\Delta S_{\text{conf}} = Nk_B\ln\frac{V_{\text{after}}}{V_{\text{before}}}. \tag{2.9}$$

---

[23]We here assume that only the volume is changed. If, for example, the energy is also changed, there are additional contributions to ΔS as will be explained later.

[24]The gas constant R, which can be defined in this way, also occurs in the ideal gas law which we shall deal with later.

Key points

- Entropy, S, is defined from the number of available microstates, Ω, by $S = k_B \ln \Omega$, where k_B is Boltzmann's constant.

- While Ω is multiplicative, $\Omega_{AB} = \Omega_A \times \Omega_B$, the entropy S is additive $S_{AB} = S_A + S_B$.

- When the volume of an ideal gas is increased from V_{before} to V_{after}, the configurational entropy increases by the amount

$$\Delta S_{\text{conf}} = N k_B \ln(V_{\text{after}}/V_{\text{before}}) = n R \ln(V_{\text{after}}/V_{\text{before}}),$$

where N is the number of particles, n the number of moles and R is the universal gas constant ($R = N_{\text{Av}} k_B$).

2.5 Gas diffusion
Mixing gases

Let us now return to our example where two gases are mixed. We divide the system into cells (see Figure 2.16). There are 32 black and 32 white particles on either side of the partition. If the white particles have Ω_{white} configurations and the black ones have Ω_{black}, this means there are a total of $\Omega_{\text{white}} \times \Omega_{\text{black}}$ configurations for the entire system. When we remove the wall, the available volume is doubled for both the black and the white particles, $V_{\text{after}}/V_{\text{before}} = 2$. This means that the number of configurations for the whites increases by a factor of 2^{32} and for the blacks also by a factor of 2^{32}. The number of configurations for the entire system therefore increases by a factor of $2^{32} \times 2^{32} \approx 2 \cdot 10^{19}$.

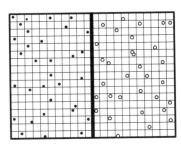

Figure 2.16   The system in Figure 2.4a is divided into imagined cells. The partition divides the box into two equal halves.

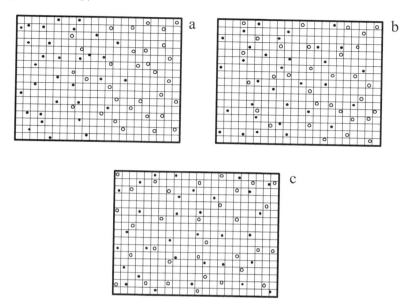

Figure 2.17 The two particle types spread throughout the entire volume in Figure (a) → (c).

We will see a process where the particles are spread over the entire volume (Figure 2.17a,b), and finally the particles are evenly distributed (Figure 2.17c). The probability that the system would then be found in any configuration among the original ones is 1 out of $2 \cdot 10^{19}$, so it is very unlikely that the system returns to its original particle distribution. The number of configurations with particles evenly distributed throughout the volume constitutes an overwhelming majority.

Once again we see that the mere fact that Ω has radically increased, means that the process is spontaneous. Of course, the entropy $S = k_B \ln \Omega$ then also has increased. We have

$$\Delta S = k_B \ln(2 \cdot 10^{19}) \approx 44 k_B,$$

so $\Delta S / k_B$ is a rather small number (we here have a small system with only 64 particles). Note that while Ω increases by a huge amount, S increases moderately, which is a consequence of the fact that a logarithm, $\ln(x)$, is a very slowly increasing function of x for large x. The logarithm in Equation (2.6) implies that S is of a convenient magnitude.

Alternatively, we can obtain the result $44 k_B$ by using Equation (2.8). The entropy increase for each kind of particle is $\Delta S = N k_B \ln[V_{\text{after}}/V_{\text{before}}] = 32 k_B \ln 2 \approx 22 k_B$. The total entropy increase is the sum of the contributions for both kinds, that is, $44 k_B$.

Like in the gas expansion, each individual particle of the ideal gas behaves as if it were alone in the box. It has absolutely no "desire" to mix with particles of the other kind, so that the system becomes homogeneous. Each particle simply moves straight ahead in an utterly selfish manner, without caring the least about the other particles. The particles move "independently" of each other.

It is often said that the system becomes mixed because of the gain in "entropy of mixing," that is, the entropy is higher for the final state when the two particle types are mixed compared to the initial state when they were separated. Note, however, that the increase in entropy is not the *reason* why the system becomes mixed. This increase is merely the result of the fact that Ω increases when the available volume for the different particle types is increased and that the system thereby goes to a more probable macroscopic state where the gases have been mixed. The reason why the system becomes mixed is that the particles can move over the entire volume. They do not care what happens to the entropy of the system – in fact they do not even know of it because each particle of the ideal gas perceives it as if it were alone in the box, while entropy describes the entire collection of particles.

The increase in entropy during mixing is an example of what is usually called **entropy production** – that entropy is "generated" within the system. Note the relationship between the direction of the process and the production of entropy: The process simply goes in the direction that is most probable, and this fact is expressed as an entropy increase.

Key points

• Mixing of two ideal gases occurs because the macroscopic state with mixed gases is much more probable (has tremendously more microstates) than the state with separated gases. The increase in entropy (entropy production) can be seen as a result of this fact.

2.6   Dispersion of energy
Energy distribution and entropy

We have seen how particles spread over the available volume, whereby the system in principle utilizes all possible particle configurations, that is, all possible microstates of the system. Each **macroscopic state** corresponds to a large number of microstates, which are all practically equal when viewed from the macroscopic perspective (but different from the microscopic perspective). The thermodynamic equilibrium state is the macroscopic state that corresponds to the overwhelmingly largest number of microstates. It is most likely that the system "goes to" and "stays with" them, because they are so many compared to all other microstates. The spontaneous process towards equilibrium is a manifestation of the fact that the system does not stay very long in macroscopic states that are improbable (unless prevented from leaving them), but sooner or later instead goes to the most probable state, that is, the equilibrium state.

To describe the state of the system at the microscopic level, the **microstate**, it is not sufficient to specify the particle positions (the configuration), but one must, among other things, also specify their energies and, in classical physics, their velocities. The microstate of a system generally constitutes a description of the system that is as complete as possible on the microscopic level.

We shall now study how the energy of a system can be spread out (dispersed) among the various molecules and we will see that the principles that apply in this case are similar to the principles of spreading of particle positions. As we know from theory of atoms and molecules/quantum physics, the energy of an atom or a molecule is *quantized*, that is, the energy can only assume a set of discrete values. This applies to any system that is enclosed in a finite volume, such as "the particle in a box" (a particle enclosed in a box), or otherwise has some spatial limitation, such as particles in a potential well (a region of low potential energy) like electrons moving around an atomic nucleus.[25] The various states of a molecule are called **quantum states**.

The quantum state with the lowest energy is called the ground state, the next quantum state of higher energy is referred to as the first excited state et cetera. (To excite a particle means to raise its energy to a higher level.) We represent each quantum state of a particle with

---

[25]A further important example is a rotating molecule whose atoms move around on closed surfaces (spherical surfaces surrounding the center of rotation).

Figure 2.18   (a) Particle in its ground state (the lowest possible energy). (b) Particle in its second excited quantum state, which has a higher energy.

a horizontal bar and mark the present state of the particle as a thick line (Figure 2.18). To begin with, we assume for simplicity that there is only one quantum state at each energy level and that the gap between two consecutive levels is equally large everywhere.

When one has a system with many particles, it is of interest to see how the energy can be distributed between them. The different microstates of the system have different distributions of energy among the particles. In this case, when the gap between the levels are equal, one can find all possible distributions in a simple manner. One can then imagine that one has access to energy quantities of a certain amount, "energy packets," which can be transferred between the particles. This amount is equal to the energy difference between adjacent levels, that is, the size of the gap. The energy contents of a molecule can then be thought of as the number of energy packets that it holds, so a content of zero packets represents the ground state, one packet the first excited quantum state, two packets the second excited state, et cetera.

This makes it pretty easy to determine the ways in which energy can be distributed between the molecules and how many different ways there are. Basically it is just a matter of drawing all possible ways to distribute packets between particles and then to count them, just as we did when we were studying how particles could be distributed among the cells in Figures 2.8 to 2.10. However, all energy packets are exactly the same so the only thing that counts is how many packets a particle holds (that is, how much energy the particle has) – not which particular packets it has.[26]

---

[26]The problem of distributing particles of the same kind between the imaginary cells that we studied in Section 2.3, is actually of a similar nature since particles of the same kind in a gas cannot be distinguished from each other (see footnote 13).

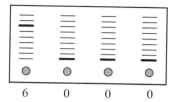

Figure 2.19   The initial state for the system with four particles.

Let us study a system of four particles that are placed next to each other.[27] We assume that they are isolated from the outside world. The first particle is initially in its sixth excited quantum state (that is, it holds "six energy packets"), while the others are in their ground state (Figure 2.19). The particles interact with each other so they will exchange energy[28] and they will continue doing this again and again as time passes. Thus particle number two can receive all the energy (all packets) from particle number one, which thereby goes down to its ground state, or it may be particle three or four that receives all of the energy, as illustrated in Figure 2.20. These four possibilities (the initial state and the other three) are four different microstates of the whole system and they are equally probable.

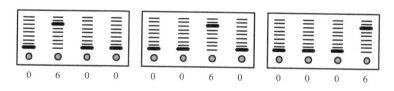

Figure 2.20   It is equally likely that any of the particles is in its sixth excited quantum state.

In addition to these four, there are many more possibilities. Each particle can exchange a portion of its energy with another particle, for instance that the energy of a particle is lowered two levels (it releases two packets) while another particle increases the energy by two levels (receiving two packets) while the remaining particles remain in their

---

[27]Thereby, we assume that they are distinguishable from each other.

[28]However, we assume for simplicity that the particles interact so weakly that they do not affect each other's energy spectrum (that is, the locations of the quantum states along the energy axis). The exchange of energy solely changes which quantum state that is occupied for each particle.

Figure 2.21   Some of the possible energy distributions where a particle is in its fourth and another in its second excited quantum state.

ground state. The first particle is thereby changed from its sixth to fourth excited quantum state, while the other goes from its ground state to its second excited state. Two of these possibilities are shown in Figure 2.21 and there are 10 more possibilities of the same kind. (EXERCISE: Draw them on a piece of paper.)

The only limitation that exists here is that the total energy U for all four particles must be the same all the time (because the system is isolated from and cannot exchange energy with the outside world). In our example, we imagine that there are six packets to distribute. Thus, for example, a particle can be in its third excited state (having three packets), another in its second (having two packets), a third one in its first (having one packet), and fourth being in its ground state. All of these possible microstates of the whole system are depicted in Figure 2.22.

If one examines all possible energy distributions for the four particles, then one finds that there are a total of 84 different combinations, that is, 84 microstates for the entire system which have the same total energy as the initial state. We thus have $\Omega = 84$. All of these microstates *are equally probable.*[29] In only four of these microstates one of the particles has all the energy, while many more microstates have energy distributed over more particles. The most numerous are those distributions where the energy is about evenly distributed over all particles.

While time passes, the system will pass from one microstate to the next and the next, and so on. Therefore, the system will leave the initial state (where all the energy is contained in particle number one) and the energy will be distributed over all molecules. The microstates where the energy is about evenly distributed over all particles are

---

[29]This follows from a fundamental postulate of statistical mechanics.

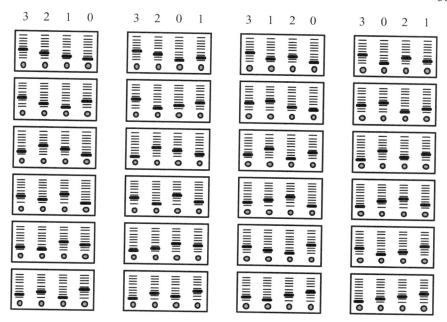

Figure 2.22   All microstates of a system where one particle is in the third, one in the second, one in the first excited quantum state, and one is in the ground state. The numbers at the top refer to the states depicted in the first line.

most numerous, so it is most likely to find the system with the energy evenly dispersed.

It is easy to see that the number of possible microstates increases rapidly with the number of particles, since more particles means a lot more ways to distribute the energy (provided that there is enough energy to distribute). For a system with many particles, the microstates with evenly distributed energy are considerably more numerous than the states where a significant portion of the energy is concentrated to one or a few particles. For a macroscopic system, the number of microstates with uniformly distributed energy is enormously large compared to the other microstates.

To make it easier for the reader to understand the consequences of these circumstances, a popular-science analogy is presented in Appendix A. The main purpose of this analogy is to illustrate spontaneous dispersion of energy and to prepare for the discussion of the concept of temperature as treated in the next section. The metaphor used in the appendix is that one has coins (corresponding to our energy packets) that are spread among humans (corresponding to particles) through their exchange of money between themselves (corresponding to the energy exchange upon interaction). The reader is recommended to study Appendix A, in particular parts A.1 and A.2, at this point.

For a molecular system where we have excited (raised) one or a few molecules to higher quantum states, for instance by heating one end of a solid body, then eventually the added energy will spread to all molecules. Since the number of microstates with uniformly distributed energy are immensely more numerous than the states where the energy is contained in a few molecules, it is very unlikely that the system will return to the initial condition. Thus we will observe a spontaneous process, where the system goes from an initial condition with small value of Ω to a final macroscopic state with a large value of Ω, which means that $S = k_B \ln \Omega$ will increase. The added energy will spread from the heated end and will become uniformly distributed over the entire system. This is an example of **heat conduction**.

Note the similarity to the distribution of particle positions over the available volume, which we discussed earlier. The most likely scenario was that the particle positions became evenly distributed, since such configurations (microstates) are by far the most numerous ones. The common denominator for particle spreading and energy dispersion is that we will observe that the system is changing in the direction that is most probable, that is, in the direction of increasing Ω and therefore increasing S. Although the physical processes are quite different – the distribution of particles over volume and distribution of energy over particles – the entropy plays exactly the same role. Entropy is thus an important concept, which in a handy manner *summarizes what is common for the two types of processes.*

In our simple system of four particles, we had 84 different microstates and we noted that the number of states increases rapidly with the number of particles (provided there is sufficient energy to distribute). Furthermore, the number of possible states is significantly greater if we have more than one quantum state at each energy level, which is the normal situation (for example we have three p orbitals with the same energy in the hydrogen atom – or six if we also consider the electron spin). We can realize this from Figure 2.23, where we show examples of some microstates for a system with four particles. We have chosen to show four of the microstates which differ since

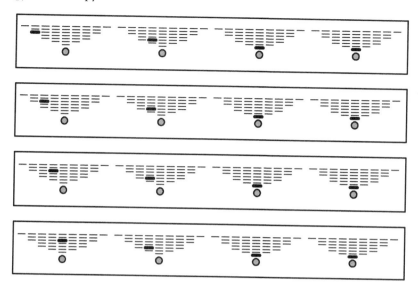

Figure 2.23 A more realistic example in which each particle has several quantum states at each energy level. Each particle can be in any of its quantum states in an energy level, which are considered as distinct possibilities that all count. Here we show just four of the possibilities – they have the first particle in different quantum states at the fourth energy level above the ground state.

particle number one is in different quantum states with the same energy (in the fourth energy level above the ground state). Moreover, molecule two can be in any of its quantum states on the second energy level, and so on. All these microstates correspond to a single state in our previous case, where each molecule had only one quantum state per energy level. One can see that the number of combinations increases a lot compared to the previous case.

Another difference between our example and actual molecules is that the gap between adjacent energy levels of the molecules is not equal everywhere, but instead the gap varies from energy level to energy level.[30] This means that the amount of energy that is available for

---

[30]Our example has nevertheless physical relevance since it corresponds to the energy levels of a so-called harmonic oscillator, such as a particle oscillating in a parabolic potential well or the vibration of two covalently bonded atoms at low energies (near the bottom of its potential well, which to a good approximation is parabolic there).

the molecules to exchange no longer can be illustrated by a number of equal "energy packets," but the exchanged energy will rather come in a variety of sizes. This makes it more complicated to figure out the number of different distributions of energy between the molecules, but the principles are the same and the consequences as well. In this book, therefore, we will not worry about these details.[31]

It is easy to see that the number of microstates becomes larger if we increase the energy of the system by exciting it to higher energy levels – it is simply so that the number of possibilities to distribute the energy becomes larger if we have more to distribute. Thus, Ω increases when U increases,

$$\frac{d\Omega}{dU} > 0 \qquad\qquad\qquad (2.10)$$

and therefore $S = k_B \ln \Omega$ increases, that is, we have[32]

$$\frac{dS}{dU} > 0. \qquad\qquad\qquad (2.11)$$

As we saw earlier, energy can be added to or removed from one part of a system to another by heat conduction. This means that one part of the system is excited to higher energy levels, while the other part is de-excited (decreased) to lower ones. If the total entropy then increases, this is a spontaneous process. The same applies to energy transfer by conduction between two different systems. To add energy to a system by exciting it to higher energy levels (without changing anything else) is an example of what is called addition of "heat." The amount of energy transferred to a system then is called **heat**. As will be discussed in the next section, one can add energy in the form of heat by bringing the system into contact with a warmer body, such as a hotplate. Similarly, one can transfer heat from a system by bringing it in contact with a colder object. In Chapter 4 of the book, we shall see that there are ways to change the energy of a system without transfer of heat, for example by doing work that is performed by compressing or expanding the system (a reduction or an increase of its volume).[33]

---

[31] The case of a monatomic ideal gas is, however, treated in detail in Appendix E.

[32] The volume and the number of particles are here assumed to be constant, so in a mathematical sense we actually have $(\partial S/\partial U)_{V,N} > 0$, where the subscript indicates which variables are held constant during the partial differentiation.

[33] When the volume of the system is changed the energy levels are changed, which does not happen during pure heat transfer when only excitations and de-excitations occur.

We will return to heat and work in Section 4.2, where we give the complete definitions of these concepts.

Key points

- Dispersion of energy (added as heat) in a body happens because there exist tremendously many more ways to distribute energy over the quantum states of the molecules when the energy is evenly spread over all molecules compared to when the energy is unevenly distributed over the molecules (that is, there are many more microstates of the entire system in the former case compared to the latter). The spread of energy accordingly occurs in the direction that is most probable and that corresponds to an increase in entropy.

- The number of ways to distribute the energy (number of microstates Ω) increases rapidly as the energy of a system increases. We accordingly have $d\Omega/dU > 0$ and hence $dS/dU > 0$.

2.7 Hotter and colder
The concept of temperature; the second and third laws of thermodynamics

We saw earlier that energy which is supplied to a part of a system, is spread over all molecules in the entire system. It is very unlikely that the energy would remain in the part of the system where it was added. The spread of energy over all molecules is much more probable. As we have seen, the reason for this is that the number of microstates with uniformly distributed energy over all molecules is *much* larger than the number of microstates with unevenly distributed energy. In the macroscopic (thermodynamic) equilibrium state, which is the most likely macroscopic state with an overwhelming majority of microstates, the energy is thus evenly distributed.

In our discussion, we assumed (without saying it) that all molecules were of the same kind. The question now is what happens if we have molecules of different kinds, for example if we have a body A and a body B of different materials and we bring them together, so that energy can be transferred as heat between them.[34] Will the energy be evenly distributed over all molecules of A and B at equilibrium? The answer is generally no. What then is it that characterizes equilibrium? From our everyday experience we know that the *temperature* of A and B are equalized, so that both bodies have the same temperature at equilibrium (thermal equilibrium). The energy per molecule in A is, however, generally different than in B; for example, molecules of a gas do not have the same energy as the molecules in a solid body even if the temperature is the same. But, then, what is temperature? And what are its properties?

If we, when we bring systems A and B together, get a spontaneous heat transfer from A to B, we say that A has a *higher temperature* than B, while A has a lower temperature than B if the reverse occurs. If we do not get any spontaneous transfer of heat, we say that the bodies have the *same temperature* (note that we are talking about a net transfer of heat; at equilibrium the heat flow from B to A is as large as from A to B). Without specifying what temperature really is, we can use these considerations as a definition of what we mean by higher, lower, and equal temperatures.

Our reasoning still applies, namely that the entire system (AB) at equilibrium has a macroscopic energy distribution that corresponds to by far the greatest number of microstates. Such an energy distribution is obviously the most likely, which means that a system develops in the direction of an increased Ω and thus increased entropy,[35] $S = k_B \ln \Omega$. Equilibrium in the combined system AB thus corresponds to the macroscopic energy distribution that provides the greatest number of microstates Ω_{AB} and hence maximum entropy S_{AB}. (We assume that system AB is isolated from the rest of the world,

---

[34]For both A and B it is assumed that the volume and the number of particles are constant during the energy transfer.

[35]The number of microstates for the macroscopic energy distribution corresponding to equilibrium is enormously greater than that of all other energy distributions. Although Ω is actually the total number of microstates for all possible energy distributions, one can therefore entirely disregard all macroscopic distributions except the equilibrium distribution in the calculation of $S = k_B \ln \Omega$ for the final macroscopic state of the system. Furthermore, note that the initial macroscopic energy distribution between A and B (that is, the distribution before we bring A and B in contact) is still possible when we allow heat transfer between A and B – it is one of the possibilities for AB. However, the number of microstates belonging to this initial distribution is so small compared to the final equilibrium one, that the probability for this possibility is entirely negligible.

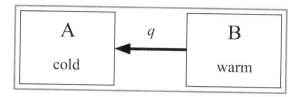

Figure 2.24 Spontaneous heat transfer between two systems A and B. The system that delivers heat to the other has the highest temperature by definition.

so that energy exchange only occurs between A and B.) We shall now examine what this leads to. The reader is advised to read the whole of Appendix A including part A.3 as a preparation.

Let us assume that body A is colder than body B so that heat transfer occurs spontaneously from B to A when the bodies are brought together (Figure 2.24). We let the bodies be in contact for a short time only, so that the net amount of heat, q, transferred to A from B is small. Body A takes up energy q and B gives away the same amount. Thus, the energy change for A is positive, $\Delta U_A > 0$, while the change for B is negative, $\Delta U_B < 0$.[36] Since Ω increases as U increases ($d\Omega/dU > 0$ according to Equation (2.10)), Ω_A will increase. For B the energy is reduced and therefore Ω_B will decrease. In other words, the number of ways to distribute energy microscopically increases/decreases as the amount of energy to distribute increases/decreases.

Since the heat transfer from B to A is a spontaneous process, the entropy of the entire system, S_{AB}, must increase. When energy goes from B to A we have seen that Ω_B decreases while Ω_A increases and hence S_B decreases and S_A increases. We have $S_{AB} = S_A + S_B$ and because S_{AB} increases it follows that S_A must increase by a larger amount than what S_B decreases, that is, $|\Delta S_A| > |\Delta S_B|$.[37]

This is illustrated in Figure 2.25, which shows the entropy for systems A and B as a function of energy, $S = S(U)$. In the figure, ① is the initial state and ② is the final state (note the placement of these symbols in Figure 2.25 for systems A and B, respectively). The arrow for ΔS_A is pointing upward (increased entropy), and for ΔS_B downward

---

[36]If A receives the heat q from B, we have $\Delta U_A = q$ and $\Delta U_B = -q$.

[37]When ΔS_A is a positive number, the entropy change of the entire system $\Delta S_{AB} = \Delta S_A + \Delta S_B$ is positive provided the negative number ΔS_B has a smaller magnitude than ΔS_A.

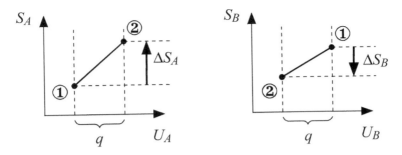

Figure 2.25   The entropy as a function of energy of the systems A and B depicted in Figure 2.24. The figure shows the change in the macroscopic state of the systems when heat is transferred spontaneously from system B to A. System A receives the amount of heat q and B gives away the same amount. System A goes from its state ① to state ②, whereby its energy increases by the amount q and its entropy increases by ΔS_A. At the same time B goes from its state ① to ② whereby its energy is reduced by the amount q and its entropy changes by ΔS_B, which is negative.

(decreased entropy). As shown in the figure, the entropy increase for A is larger than the decrease for B. Since $|q|$ is the same for A and B, we see that the slope of the curve $S = S(U)$ is greater for A than for B, that is, the derivative dS/dU is greater for A than for B

$$\frac{dS_A}{dU_A} > \frac{dS_B}{dU_B}. \tag{2.12}$$

Thus we see that the subsystem with the *lowest* temperature has the *largest* derivative dS/dU.

    If we bring the systems together briefly once more and repeat this again and again, energy (heat) is spontaneously transferred from B to A as long as the total entropy increases, $\Delta S_{AB} > 0$, that is, as long as $|\Delta S_A| > |\Delta S_B|$. This will take place until the entropy does not increase any further, whereby we have $\Delta S_{AB} = 0$ and $|\Delta S_A| = |\Delta S_B|$. Then no net energy is spontaneously transferred between the subsystems, which means that the temperature of A and B have become equal. Since $|\Delta S_A| = |\Delta S_B|$ we have then come to a macroscopic state of the system where the slope of the curve $S = S(U)$ is equal for A and B, that is,

$$\frac{dS_A}{dU_A} = \frac{dS_B}{dU_B} \qquad (2.13)$$

at equilibrium. Thus, the two systems with the same temperature have *the same value of the derivative dS/dU.*

We see that the derivative dS/dU is closely associated with temperature. If the derivative is equal for two systems, the temperature is the same, and if the derivative is different, the temperature is different – the system with the largest derivative has the lowest temperature. We could, therefore, use the value of the derivative as *a measure of temperature*. However, it is impractical to use dS/dU itself as the value of the temperature, because we want a warm body to have a higher temperature than a cold and dS/dU has a lower value for a warm body. Instead, we can take

$$T = \frac{1}{dS/dU} \qquad (2.14)$$

as a value of the temperature (this relationship defines T).[38] A hot body then has a higher value of T than a cold body and equal values of T mean that the temperature is the same. From Equation (2.11) we see that the derivative is positive, so we always have $T > 0$, which means that negative values of T cannot be obtained (for a system at equilibrium). The quantity T is called the **absolute temperature**.[39] Its unit is K (Kelvin).[40]

Conversely, we can say that the absolute temperature is a measure of how much the entropy increases in a system when we add a certain amount of heat.[41] Low temperature means that the entropy changes a lot and high temperature means that it changes a little. Therefore, heat flows spontaneously from high to low temperature;

---

[38]Since the volume and the number of particles are here assumed to be constant, we have mathematically $T = 1/(\partial S/\partial U)_{V,N}$.

[39]We shall see later (in Section 4.4) that this is exactly the same quantity as the absolute temperature of the ideal gas law.

[40]Since the unit for S is JK^{-1} and that of U is J, the derivative dS/dU has the unit K^{-1} and hence the unit of T in Equation (2.14) is K. This relationship justifies the choice of a unit for k_B (and hence for S) that is given by energy/temperature.

[41]Expressed more accurately, $1/T$ (which is equal to dS/dU) describes the rate at which S is changed when one changes the energy.

the body with low temperature gains more entropy than that with high temperature loses, whereby the total entropy increases.[42]

Another important relationship can be obtained from our reasoning. For small values of q, the slope of the curve $S = S(U)$ in Figure 2.25 is equal to $\Delta S/q$, but this slope is also equal to $dS/dU = 1/T$. Thus we have $\Delta S/q = 1/T$, that is,

$$\Delta S = \frac{q}{T}. \tag{2.15}$$

This is a special case of what is called the *second law of thermodynamics* and it relates the heat transfer to a system and the entropy increase for the latter (we assume here that q is so small that the temperature does not change because of the heat input). When we take heat from the system, q is negative and so is ΔS (the heat q is always counted as that *added* to the system, so q is negative when heat is removed).

However, we have seen that the entropy of a system can increase even in the absence of heat transfer; Ω and hence S increase when a spontaneous process occurs within the system, for example when the two gases are mixed in it. If such a process occurs at the same time as heat is added, the entropy increase will be greater than q/T. A more complete statement about ΔS therefore reads

$$\Delta S \geq \frac{q}{T}, \tag{2.16}$$

where the inequality is valid when some spontaneous process (in addition to heat transfer to or from the system) occurs in the system.

In the concrete examples we have discussed so far, we have either had spreading of non-interacting molecules within a volume or spreading of energy among weakly interacting molecules. In real systems, we often have molecules that interact strongly and that are mobile. Then we need to simultaneously consider both the positions of the molecules and their interactions (the strength of the interactions between the molecules depend on their relative distances). This complicates the treatment considerably. However, the same general principles still apply, namely that the most likely macroscopic distribution of energy and particles constitutes the thermodynamic equilibrium state. For this distribution, there are tremendously many

---

[42]An alternative derivation of the relationship between entropy and temperature can be found in the box at the end of Appendix A.

more microstates than for other distributions and hence this macroscopic state has greater entropy.

A spontaneous process is always **irreversible**, meaning that it cannot run backwards in such a manner that one returns to exactly the same macroscopic state for both the system and its surroundings. Since the total entropy has increased during the process, there is no way to reduce it again – all processes that can happen are spontaneous and give increased total entropy. If we try to make the entropy to decrease somewhere, it will increase by a greater amount elsewhere so that the total entropy nevertheless increases. (Even if we ourselves try to intervene and change the course of events, that will not help because we too are driven by increased entropy.)

A **reversible process** is a process that in principle takes place infinitely slow and where the system is always at equilibrium. Such a process does not give rise to an increase in the total entropy. It can be run backwards and one then returns to exactly the same state as one started from for both the system and the surroundings. This can never be carried out in reality – every real process is irreversible. However, one can perform processes that are almost reversible (one can in principle come as close as one likes to a reversible process). Thereby one runs the process very close to equilibrium, so that it gives rise to only a small entropy increase. This increase can be made smaller by running the process even slower and closer to equilibrium. In the limit when the process runs infinitely slow and at equilibrium, the entropy increase becomes zero. (This will be discussed in more detail in Section 4.3.)

Consider a system where irreversible processes occur at the same time as heat transfer to the system. The entropy change due to the latter is equal to q/T. Let ΔS_{irrev} denote the increase in entropy in the system due to the irreversible processes (i.e., in addition to the entropy change from the heat transfer). The total entropy change of the system is thus equal to

$$\Delta S = \frac{q}{T} + \Delta S_{\text{irrev}}. \tag{2.17}$$

Since one always has $\Delta S_{\text{irrev}} \geq 0$, we see that Equation (2.17) is another way to express Equation (2.16). For reversible processes one has $\Delta S_{\text{irrev}} = 0$ and we obtain Equation (2.15).

Note that we have assumed that q is small. To emphasize this, one should use the notation dq instead of q and dS instead of ΔS (dq and

dS represent small amounts: a small amount of heat added to the system and small increment of S for the system)[43] and we generally write

$$dS = \frac{dq}{T} + dS_{\text{irrev}} \quad \text{where } dS_{\text{irrev}} \geq 0 \qquad (2.18)$$

or, equivalently,

$$dS \geq \frac{dq}{T}. \qquad (2.19)$$

In both cases the inequality holds for irreversible processes and the equality for reversible processes. (Equations (2.15) to (2.17) are, however, applicable for heat transfers q and entropy changes ΔS that are not small, provided the temperature T is unchanged during the process.)

Equation (2.18) (or Equation (2.19)) is a complete form of the **second law of thermodynamics**, which we shall return to in more detail in Section 4.3. Expressed in words, it says that the entropy of a system can be changed by heat added to/removed from the system or by any irreversible (spontaneous) process that occurs in the system. The first contribution to dS is given by dq/T and the last, dS_{irrev}, is always positive (for irreversible processes) or zero (for reversible processes). The amount of heat dq is positive when heat is added to the system and dq is negative when it is removed. The same applies to dq/T since $T > 0$.

A *postulate* is an unproven assumption that is held as being true. The first and second laws are two of the postulates that the whole of classical thermodynamics is built up from by logical and mathematical reasoning. In our treatment, we get insights into what the second law expresses and means from a molecular perspective, something that classical thermodynamics is not concerned with (at least not in its pure classical form).

We have seen examples of how entropy plays a crucial role in spontaneous processes of a system. A process is spontaneous provided the *total* entropy increases. It is important to include all entropy changes,

---

[43]Generally, we use the symbol Δf to denote the change in a quantity f (regardless of the magnitude of the change) and the symbol df to designate a *very small* change in f (in principle an infinitesimally small change, $df \to 0$). Here f symbolizes any property of a system like entropy S, energy U, and volume V. (The quantity df is what is called the differential of f in mathematics.) The term dq refers similarly to a *very small* amount of heat added to the system, while q is the added amount of heat no matter how large it is.

so that we really obtain the total entropy change. If our system interacts with its surroundings, we must also include the change in the entropy of the surroundings, S_{surr}. It is the total entropy $S_{tot} = S_{system} + S_{surr}$ that increases during a spontaneous process. S_{system} may increase or decrease depending on the circumstances.[44] If we, for example, have a warm system in contact with a cold environment, S_{system} spontaneously decreases as heat flows from the system to the surroundings. S_{surr}, however, will increase more than S_{system} decreases, so the total entropy increases. It is commonly said that the entropy of the "whole universe" increases during spontaneous processes – which we, however, should take with a grain of salt, because we hardly know enough about the universe to say such a thing with certainty.

That entropy increases is often popularly described as a decrease in order – a disordered macroscopic state would thus be more probable than an ordered one. This is usually true but not always. There are examples of systems that have higher entropy in an ordered state than in a disordered one. Remember that high entropy corresponds to many different possibilities for the system – many different particle configurations and different energy distributions. An example from everyday life is in order. If we fill a large box with books that we just throw down in a helter-skelter manner, we find that the books lock one another's positions when the box is full. Their freedom to move around is very limited when we shake the closed box slightly. If we instead pack the box with the same books by neatly arranging the books in the box, we find that there is a rather large empty space left. When we shake the closed box in this case, the books have much more freedom to move. There are more arrangements available to them.[45] If we consider all possible arrangements of the books, including all possible helter-skelter ones and those attained during the shaking, the ordered arrangements are much more numerous. Thus, the entropy is higher in the ordered state. If one wants to make an analogy between entropy and some commonplace concept, **freedom** is better than disorder. A high entropy thus corresponds to a high freedom.

Important examples of ordered molecular systems with high entropy are liquid crystals, which are found in many dials, calculators,

---

[44]In other parts of this book we use the notation S for S_{system}, the subscript "system" is then implied and not written out.

[45]To avoid the complications due to the effects of gravity on the packing of heavy books, it is better for argument's sake that we replace books with, for instance, lightweight styrofoam discs of the same size.

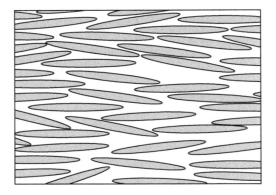

Figure 2.26   A sketch of cigar-shaped molecules in a liquid crystal. The figure shows a typical snapshot picture of the mutual orientations of the molecules. They have their longitudinal axes approximately parallel and exhibit a relatively orderly phase with higher entropy than the corresponding disordered phase where the axes are more randomly oriented.

and monitors. The molecules of a liquid crystal are in most cases elongated ("cigar-shaped") and they spontaneously form ordered liquid phases, where a snapshot picture of the structure may for example look like Figure 2.26. This ordered phase has higher entropy than a disordered phase where the longitudinal axes of the molecules are no longer approximately parallel. The reason for this is basically the same as in our example with the books.

We shall conclude this section by examining the entropy at low temperatures. When we reduce the energy of a system, its temperature gradually decreases.[46] Eventually, the system will approach its quantum mechanical ground state. At the ground state the temperature $T = 0$, the absolute zero. For most pure substances in crystalline form (perfect crystal), there is a single ground state, that is, $\Omega_{ground} = 1$ (one single way to distribute energy and particle positions in the crystal). Entropy is according to Equation (2.6) at absolute zero

$$S_{T=0} = k_B \ln \Omega_{ground}.$$

When $\Omega_{ground} = 1$, this implies that $S_{T=0} = 0$ since $\ln 1 = 0$.

---

[46]At phase transitions such as condensation (gas-liquid phase transition) and freezing (liquid-solid phase transition), the temperature will, however, remain constant until all of the substance has been transformed from one phase to the other. Such transitions are discussed in Section 6.1.

For pure substances we accordingly have that $S \geq 0$ at absolute zero, where the equal sign applies to substances in perfect crystalline form. This statement is called the **third law of thermodynamics** and it is a postulate in classical thermodynamics.

Key points

- The absolute temperature is a measure of how much the entropy increases in the system when we add a certain amount of heat. $1/T$ (which is dS/dU) indicates the rate of the entropy increase when energy is increased.

- Heat is transferred spontaneously from a hot to a cold body since the body with low temperature gains more entropy than the body with high temperature loses.

- The **second law of thermodynamics** can be formulated

$$dS = \frac{dq}{T} + dS_{irrev},$$

where dq is the added heat, T is absolute temperature and $dS_{irrev} \geq 0$. If any irreversible (spontaneous) process occurs apart from heat transfer, $dS_{irrev} > 0$. Otherwise $dS_{irrev} = 0$, that is, $dS = dq/T$.

- The most probable macroscopic distribution of energy and particles (for the system and its surroundings) is the thermodynamic equilibrium state. For this distribution there are tremendously many more microstates than for other distributions and hence this macroscopic state has largest total entropy.

- The total entropy of a system and its surroundings, $S_{tot} = S + S_{surr}$, increases in a spontaneous process.

- High entropy can be likened to high freedom. A disordered system often has a higher entropy than an ordered one, but this is not always true.

- The **third law of thermodynamics** states that for pure substances $S \geq 0$ at absolute zero, $T = 0$, whereby the equality sign applies to substances in perfect crystalline form.

2.8   Availability of energy*[47]
The Boltzmann factor

We have previously obtained the following important relationships between entropy, energy, and absolute temperature (Equations (2.6) and (2.14))

$$S(U) = k_B \ln \Omega(U) \quad \text{and} \quad \frac{dS}{dU} = \frac{1}{T}. \tag{2.20}$$

When we insert the first equation into the second, we obtain

$$\frac{d \ln \Omega(U)}{dU} = \frac{1}{k_B T}. \tag{2.21}$$

According to the definition of a derivative of a function $f(x)$, we have

$$\frac{f(x) - f(x - \Delta x)}{\Delta x} \rightarrow \frac{df(x)}{dx} \quad \text{when } \Delta x \rightarrow 0.$$

Applied to the left-hand side of Equation (2.21) with $f = \ln \Omega$ and $x = U$ this becomes

$$\frac{\ln \Omega(U) - \ln \Omega(U - \Delta U)}{\Delta U} = \frac{1}{k_B T}, \tag{2.22}$$

which is applicable for small ΔU (in the limit $\Delta U \rightarrow 0$). We can also apply this equation if ΔU is so small relative to U that the temperature T does not change when we change energy with the amount ΔU.

By multiplying both sides of Equation (2.22) with ΔU and rearranging the terms we obtain

$$\ln \Omega(U - \Delta U) = \ln \Omega(U) - \frac{\Delta U}{k_B T}, \tag{2.23}$$

which we can write as

$$\Omega(U - \Delta U) = \Omega(U) e^{-\frac{\Delta U}{k_B T}}. \tag{2.24}$$

Here, $\Omega(U)$ is the number of microstates of a system with energy U and $\Omega(U - \Delta U)$ is the number of such states after the system has lost energy ΔU. We can thus conclude that the number of microstates for

---

[47]Items marked with an asterisk (*) are optional specialized topics.

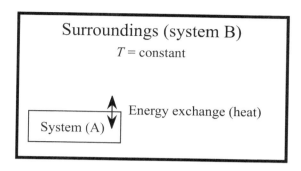

Figure 2.27   A system (A) that has a free heat exchange with its surroundings (B), which has temperature T. If B is much greater than A, the temperature will not be appreciably affected by the exchange of heat and remains constant. At thermal equilibrium A also has temperature T.

a system changes with a factor $\exp(-\Delta U/k_B T)$ when the system gives off energy ΔU at constant temperature.[48]

Let us apply this to a very large system B, which is in thermal contact with our system A (i.e., heat can be freely exchanged). The system B may be the surroundings of our system and we assume that it has a constant temperature, as depicted in Figure 2.27. We also assume that A and B are in thermal equilibrium, i.e., they have the same temperature. This means that the temperature of A is constant too when heat is exchanged.

When system B gives off energy ΔU to A, the number of available microstates Ω_B will accordingly decrease, which is not favorable for B. The larger ΔU, the smaller $\Omega_B(U - \Delta U)$ according to Equation (2.24). The exponential function is rapidly decreasing for a negative argument, so the reduction of Ω_B becomes more and more noticeable when ΔU increases. This means that it is increasingly unlikely that system B gives off an energy amount ΔU when this amount becomes larger. Thus, it is unlikely that system A can obtain large amounts of energy from B. We can express this in the following manner: The factor $\exp(-\Delta U/k_B T)$, which is called the **Boltzmann factor**, is a measure of the availability of energy ΔU from the surroundings at temperature T.

---

[48]We use the notation ΔU here since we will apply these results to another system that *receives* the amount ΔU.

Energy contributions smaller than $k_B T$ are fairly easy to obtain for a molecule from its environment, while contributions appreciably greater than $k_B T$ are hardly accessible since $\exp(-\Delta U/k_B T)$ then is very small. Since $k_B T$ is proportional to T, we see that larger energy contributions are available at a higher temperature than at lower ones.

One can take this argument one step further and thereby obtain an expression for the probability that system A is in a particular microstate with energy U_A at constant T. This expression is an important result in statistical mechanics, called **Boltzmann's distribution law** that is treated in Appendix B.[49]

Key points*

- The function $\exp(-\Delta U/k_B T)$ is a measure of the availability of energy contributions ΔU from the surroundings at temperature T.

- Energy contributions smaller than $k_B T$ are fairly easy to obtain for a molecule from its environment, while contributions appreciably greater than $k_B T$ are hardly available.

---

[49]The details including the derivation are given in Appendix B. Here it suffices to mention that the probability for system A to be in a particular microstate with energy U_A is proportional to $\exp(-U_A/k_B T)$ at temperature T (as shown in Equations B.1 and B.3). The appendix also shows how this distribution law can be applied to give the distribution of molecular speeds used in the construction of Figure 2.1.

CHAPTER 3

Entropy and free energy

3.1 Poorly soluble substance
Particle locations and energy

Let us fill a beaker with solvent and add a spoonful of any poorly soluble, crystalline substance, which settles in a pile on the bottom (Figure 3.1). Irrespective of how poorly soluble the substance is, after a while there will always be a number of molecules of the substance in solution.[1] Why? For a molecule on a crystal surface in contact with the solvent, the probability is never zero that it will detach, so sooner or later this will happen and the released molecule will "wander out" into the solvent. As we have seen in Section 2.1 there are always a few molecules with significantly higher energy than the others, and they can therefore detach from the crystal.

Figure 3.1 A poorly soluble substance at the bottom of a beaker filled with solvent. The white dots represent solute molecules that are tremendously exaggerated in size.

Once out in the liquid, the molecule will move freely over the entire volume, and will remain in solution until it eventually happens to come back to some crystal on the bottom where it can again

---

[1]If the volume is small and the substance is extremely poorly soluble, there will be solute molecules in solution at least part of the time.

attach. (We can in this case ignore the possibility that the molecule encounters a number of other dissolved molecules and combines with them to form a new crystal.)

It will of course on average take considerable time before the molecule comes back and attaches to some surface[2] and during this time, a number of other molecules will have become detached from the crystal surfaces. The number of molecules that come off from the crystal surfaces during this time interval determines how high the concentration becomes at equilibrium. If many detach, the concentration becomes higher than if few do so. At equilibrium, there is an equal number of molecules that come off per unit time as those that come back and attach.

The reason why a released molecule wanders around in the liquid is, of course, that nothing prevents it from doing so. Since the released molecules can wander around, the system acquires many microstates with various particle configurations. This means that the system obtains a positive entropy contribution when molecules are released. It is customary to express this as: *the system gains entropy of mixing*. We must remember, however, that this only means that the dissolved molecules wander around randomly in the available volume, like in the mixture of two gases discussed earlier. The equilibrium concentration is low – despite this gain in entropy – because for poorly soluble substances the probability is small that molecules detach from the crystals. What, then, is the reason for this low probability? Why are some substances poorly soluble in certain solvents?

There are many reasons why various substances are poorly soluble. Here we will examine one possible mechanism and for the sake of it, we make a simple model of our system. We let the crystals at the bottom be represented by a rectangular (prism-shaped) body as shown in Figure 3.2, and we shall examine what can happen when a solute molecule approaches the surface of the body (the solid phase).

In the example that we will consider, the energy is much lower when the molecules of the poorly soluble substance are in contact with the solid phase (of the same substance) than when they are surrounded by solvent molecules. This may be due to very strong

---

[2]Not all molecules that come back will attach immediately, but many of them do. Some molecules will return into the solution and come back later to some crystal surface.

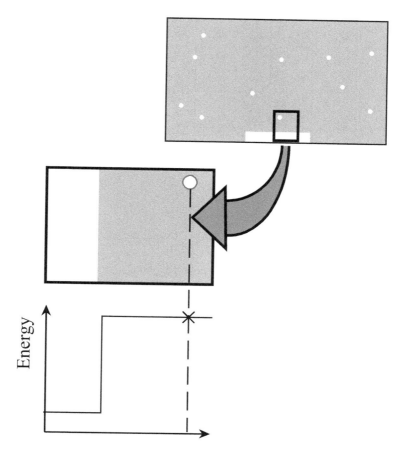

Figure 3.2 The upper figure shows a poorly soluble solid (white rectangle) in contact with a solvent (gray region). The white circles are solute molecules. The middle figure shows a magnified view of the framed area in the upper figure. This figure is rotated 90 degrees relative to the upper. In this example the energy is lower when the molecules of the substance are in the solid phase than when they are surrounded by solvent molecules. This is represented by the curve in the plot labeled "Energy." The energy has a low value to the left compared to the value at the place where the molecule is located in the figure (marked with a cross on the energy curve). The region where the energy is low is called a potential well.

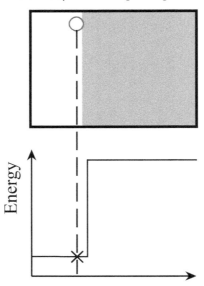

Figure 3.3   The molecule binds to the solid phase – it "goes down" into the potential well – whereby energy is released.

binding between the molecules in the solid phase.[3] When a dissolved molecule comes in contact with the surface, energy is released when the molecule becomes bound to the solid phase, as Figure 3.3 illustrates. This energy will be spread over the whole system (similar to the heat dispersion we described earlier).

By the spreading of the released energy, a huge number of microstates with various energy distributions becomes available and a positive entropy contribution is obtained. When the released energy is large, as in this case, this entropy contribution will be large. The choice for the system is thus between having many particle configurations when molecules are in solution or having many energy distributions when the molecules are bound. Since the binding to the solid is very strong (the potential well is deep), the highest number of microstates is obtained in the latter case, i.e., when many particles are bound and only a few are in solution. Therefore, this condition is most probable and the equilibrium concentration is therefore low.

---

[3]In this example, the binding energy between one molecule of the poorly soluble substance and other molecules of the same kind is much greater in magnitude than the interaction energy between the molecule and the solvent molecules and between the solvent molecules in the absence of the solute. Therefore the first-mentioned binding energy dominates in the total energy change upon binding to the solid phase.

What happens when a molecule detaches from the crystal surface? What is required for it to do so? A molecule must get out of the potential well in order to be detached from the surface, and this requires that energy is supplied. The probability for a molecule to come off therefore depends on the availability of energy from its environment and on the depth of the potential well. The deeper the well, the more energy is needed for the molecule to be released and one can show that the availability of large amounts of energy is small from the surroundings.[4] Since the potential well is deep, there is only a small probability for the molecule to detach, leading to the low solubility. If the temperature is increased, large energy contributions become more accessible and the solubility of the substance will increase. (At high temperatures there is a lot of energy to distribute between all molecules, and therefore it is natural that even relatively large energy contributions become more easily accessible.)

In this example we see how low energy (in the form of energy of binding) and high entropy (in the form of entropy of mixing) are opposed. Here entropy and energy are weighed against each other and the optimal condition consists of a compromise between these two factors. This trade-off determines the equilibrium state, which is the most probable macroscopic state of the system. Both factors can alternatively be expressed as the condition that the total entropy (from both particle configurations and energy distributions) for the system and its surroundings is maximal at equilibrium. This will be further explored in what follows.

3.2   Evaporation of a liquid drop
Balance between entropy and energy; vapor pressure

Let us put a drop of liquid at the bottom of a small closed box filled with, for example, nitrogen gas (Figure 3.4). What will happen? A number of liquid molecules will leave the drop and go out into the gas phase. In other words, at least part of the liquid will evaporate (vaporize). The gas in the box will be a mixture of nitrogen and liquid vapor. The gas pressure that is initially equal to the nitrogen gas

---

[4]In Section 2.8 it was shown that the availability of an energy amount ΔU decreases rapidly with increased ΔU. Energy contributions smaller than $k_B T$ are fairly easy to obtain for a molecule, while contributions appreciably greater than a few $k_B T$ are much less available.

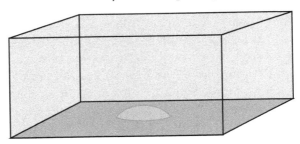

Figure 3.4   A closed box with a drop of liquid on the bottom. The box is filled with nitrogen gas that does not dissolve appreciably in the liquid.

pressure will then be given by the combined pressure of the nitrogen gas and the vapor. The contribution to the gas pressure from each molecular species is called **partial pressure**[5] whereby the total pressure is the sum of the partial pressures of all molecular species present in the gas.

What happens near the surface of the liquid during the evaporation (Figure 3.5)? For a molecule to be able to leave the liquid phase it is required that sufficient energy is available. Why? In the gas phase a molecule interacts very weakly with other molecules, so the interaction energy is near zero. A molecule in the liquid phase, on the other hand, interacts quite strongly with the surrounding molecules – it is in a kind of potential well. The interaction between our molecule and the surrounding molecules is mainly attractive (see Figure 3.6), so the potential energy is negative in the well. This is the reason why energy must be supplied to the molecule; otherwise it is unable to leave its potential well in the liquid and go out into the gas phase. This energy must be taken from the molecule's environment and, as in our previous example, the availability of energy is crucial.

The fact that few molecules have high energy, as discussed in Section 2.1, has to do with the fact that large energy contributions are available only to a small extent. The greater the energy contribution that is required, the less likely it is that it is available.[6] If the interactions between the liquid molecules are strongly attractive, the potential well for each molecule is deep and the probability is relatively low that sufficient energy is available, so rather few molecules per unit time will be able to leave the liquid phase. If the interaction between

[5]The concept of partial pressure is formally defined in Section 5.1.
[6]This is discussed in Section 2.8.

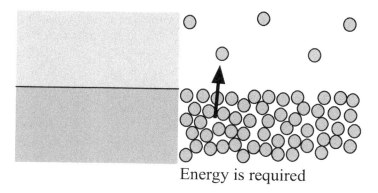

Energy is required

Figure 3.5 For a molecule to be able to leave the liquid phase, energy is required.

the molecules is not so large, less energy is required to enable them to leave the liquid and the probability is greater that this amount of energy is available. Therefore, more molecules will leave the liquid per unit time.

When a molecule has gone out into the gas phase, it will move freely throughout the available volume of gas and it will take some time until it by chance returns to the liquid drop. During the time it takes on average for a molecule to return to the liquid phase, a number of additional molecules will have left the liquid. The number of molecules in the gas phase and hence the partial pressure of the vapor will therefore increase at the beginning. The larger the amount of molecules in gas phase, the more often some molecule will return

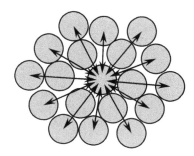

Figure 3.6 Attractive interactions between molecules in liquid. This attraction keeps the molecules together as a liquid.

to liquid phase. The number of gas molecules continues to increase until equilibrium is reached – equilibrium takes place when as many molecules leave the liquid per unit time as come back (we assume that the drop is large enough not to evaporate completely before equilibrium is attained). The value of the partial pressure of the vapor at equilibrium is called the **vapor pressure** of the liquid.[7]

For a liquid with weak intermolecular attractions, the partial pressure at equilibrium is high since the probability that molecules will leave the liquid phase is high, while the partial pressure will be lower if the attraction is strong. A concrete example of a liquid with weak intermolecular interactions is common ether (diethyl ether) and one with strong interactions is water. If we increase the temperature, the availability of energy will be higher, which leads to an increased probability that molecules will leave the liquid phase. Thus, the vapor pressure increases with increasing temperature, more quickly for ether than for water.

The system will get access to a larger number of particle configurations when molecules go out into the gas phase, and thereby the configurational entropy is increased. On the other hand, when molecules leave the gas phase and re-enter the liquid phase, energy is released, which can spread throughout the surroundings. More microstates become available when the energy is released and distributed over many molecules, i.e., there is a positive entropy contribution. The two alternatives, both of which provide increased entropy, oppose each other and the equilibrium state is a compromise between them, where the total entropy is greatest. As in the previous example, we can alternatively describe this as a compromise between low energy (intermolecular attractions in the liquid) and high configurational entropy (number of available particle configurations).

If the volume is so great that the drop evaporates completely before equilibrium is reached, the partial pressure of the substance becomes lower than the vapor pressure of the liquid. For moist air, one uses the concept of **relative humidity**. A relative humidity of x% means that the partial pressure is x% of the vapor pressure at the temperature in question. At 50% relative humidity the partial pressure of

---

[7]For a pure substance (one-component system) the vapor pressure is the pressure of the gas phase in equilibrium with the liquid. In the presence of another substance, in this case nitrogen gas, we assume that its presence does not affect the equilibrium. This is close to the truth if the solubility of this other substance is low in the liquid phase and the gas phase is approximately ideal.

water is half of the vapor pressure of liquid water, and at 100% relative humidity the water vapor is in equilibrium with water drops.
If we increase the temperature of a gas with a constant partial pressure of water vapor, the *relative* humidity of the gas will *decrease* since
the vapor pressure increases with increasing temperature (the partial
pressure becomes a smaller fraction of the vapor pressure since the
former is constant while the latter increases). Conversely, a mass of
air that has low relative humidity at a high temperature will have a
higher humidity at a low temperature. If the partial pressure of water
is higher than the vapor pressure, water will condense until the partial pressure equals the vapor pressure. This is what happens when
fog (tiny water droplets) forms when moist air cools in the evening.
Thus the vapor pressure is the "saturation pressure," i.e, the maximal
partial pressure of the vapor for a given temperature.

We can also understand this condensation phenomenon upon
cooling by the following reasoning. For liquid droplets to be formed
spontaneously in the cooled gas, the total entropy must increase. A
drop is formed because the molecules are attracted to each other when
they come sufficiently close together.[8] The molecules "go down" into
potential wells because of this attraction and thereby they release energy that will disperse into the surroundings. This energy dispersion
leads to an increased entropy. If this entropy increase is greater than
the loss of configurational entropy when the molecules leave the gas
phase, condensation will occur spontaneously. Because dispersion of
a certain amount of energy leads to a larger entropy increase at low
temperatures than at higher temperatures (according to our findings
in Section 2.7),[9] condensation will take place when the temperature
is low enough. (If the temperature is very low, ice [snow] is formed
instead of liquid.)

Key points

- A system gains in entropy when a particle becomes bound in a
  potential well, because the released energy can be distributed in a
  great variety of ways throughout the system and the surroundings.

---

[8]In practice, the initial formation of droplets during condensation usually occurs
when the vapor molecules come together on some small particles that happen to be
in the gas, like grains of dust.

[9]See Equation (2.15) applied to the surroundings with q being the energy that is
dispersed. For a given q, ΔS becomes larger when T is decreased.

This makes it advantageous to bind many particles provided the potential wells are sufficiently deep. At the same time the configurational entropy of the particles decreases.

• To lift a particle from a potential well, an energy contribution is needed from its environment. The probability that a particle leaves a deep well is smaller than it leaving a shallow well since large energy contributions from the surroundings are considerably less available than smaller ones.

• The availability of energy from the surroundings increases with increasing temperature. Therefore, the probability for particles to be lifted out of potential wells increases when temperature is increased.

• When a particle leaves a potential well and becomes free, it takes on average a certain amount of time before it binds again because it can wander around randomly in the available free volume. The average number of particles that become free during this time determines the equilibrium concentration in the solution or the partial pressure of the vapor phase, respectively.

• If the particles sit in deep potential wells, relatively few particles per unit time will become free and the equilibrium concentration or partial pressure becomes low.

3.3   Combustion of magnesium
Exothermic reaction with loss of S_{conf}

We have all seen burning magnesium in, for example, fireworks. Magnesium burns with an intense white flame in air while forming magnesium oxide

$$2\,Mg(s) + O_2(g) \; \rightarrow \; 2\,MgO(s),$$

where (s) and (g) stand for "solid" and "gas," respectively. What is it that drives this reaction forward? The reaction is highly exothermic ("heat releasing") – hence the intense flame. A considerable part of the chemical energy stored in the reactants[10] is released when the product is formed (see Figure 3.7).

---

[10]The energy is "stored" as potential energy in the interaction between the particles that the reactants consist of (electrons and atomic nuclei) and kinetic energy of these particles.

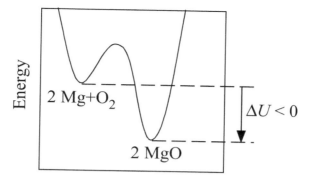

Figure 3.7   Schematic energy diagram for the reaction $2\,Mg(s) + O_2(g) \rightarrow$ $2\,MgO(s)$. Magnesium can combine with oxygen after ignition, whereby the energy barrier is overcome, i.e., the energy maximum between the reactants and the products. The resulting magnesium oxide has a lower energy; the energy difference between the product and the reactants is ΔU, which is negative. The released energy is used partially to ignite further Mg whereby the reaction proceeds.

Is it the energy reduction by the amount ΔU that drives the reaction forward? Well, not in itself – the crucial matter is what happens to the released energy. It will spread in the surroundings, whereby many molecules will become excited to higher energy levels, including higher translational, rotational, and vibrational levels – gas molecules will move more rapidly and rotate faster, and all molecules will vibrate more strongly (in short, the surroundings heat up). Even some electronic states will be involved, and light is emitted. Energy is distributed in all possible ways. The system and the surroundings gain access to a large amount of microstates, which were not available when the energy was stored in the reactants. Thus, the entropy increases overall. The probability that the dispersed energy will spontaneously gather again and make the reaction reverse is vanishingly small. The similarity to our example with heat dispersion is evident.

In the reaction, gaseous oxygen is bound and the product is a solid body (Figure 3.8). Consequently, the number of possible particle configurations is reduced substantially, so there is a negative entropy contribution which is unfavorable. The released energy is, however, so large that the entropy increase due to its spreading dominates strongly and the total entropy increases. We can say that it is this entropy increase that drives the reaction forward. However, what really

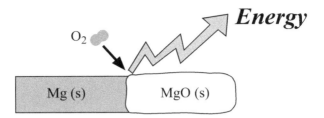

Figure 3.8   During the combustion of magnesium, oxygen is bound and magnesium oxide (MgO) is formed. Thereby the energy, that was stored in the reactants before the reaction, is released.

matters is that it is much more likely for the energy to spread over the entire system rather than coming back and making MgO decompose into Mg and O_2.

3.4   Burning candle
Exothermic reaction with gain in S_{conf}

Another example of an exothermic reaction is the combustion of a candle made, for example, of stearin (here assumed to be pure stearic acid,[11] as in Figure 3.9). Stearic acid reacts with oxygen to form carbon dioxide and water:

$$CH_3(CH_2)_{16}COOH(s) + 26\ O_2(g) \rightarrow 18\ CO_2(g) + 18\ H_2O(g).$$

A large part of the chemical energy of the reactants (stored, inter alia, in their chemical bonds) is released in the reaction. Also in this case it is essential that the released energy is spread in the surroundings, whereby the entropy is increased. However, there is another beneficial contribution. From the reaction formula, we see that 26 molecules of oxygen are consumed while 18 carbon dioxide and 18 water molecules are formed, i.e., an increase of 10 gas molecules. This means that for each stearic acid molecule that is combusted (Figure 3.10), there arise 10 additional gas molecules that can move freely in the available volume. Therefore, the number of available particle configurations increases, so the entropy increases for this reason too.

It is this spreading of energy and particles that drives the reaction forward. It is extremely unlikely that the reverse would happen, i.e.,

---

[11]"Stearin" can denote stearic acid or a triglyceride of stearic acid (tristearin). In commercial products other fatty acids may also be involved.

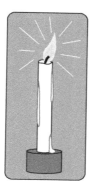

Figure 3.9   A candle containing stearic acid.

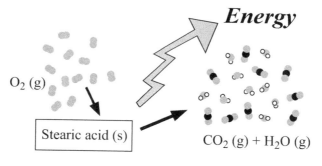

Figure 3.10   During the combustion of stearic acid, energy is released that initially was stored in the reactants. In the reaction, 26 oxygen molecules are consumed per stearic acid molecule and 36 gas molecules are formed (half of them are shown in the figure).

that carbon dioxide and water molecules would gather at the same place (and in the correct orientation) and at the same time as sufficient energy gathers there, so the reaction would have any semblance of a chance to go the other way.

Key points

- The energy released from the reactants during an exothermic combustion is distributed over all molecules in the system and the surroundings. This drives the reaction forward and the entropy increases. It is unlikely for energy to be concentrated spontaneously, which is one of several necessary conditions for the reaction to be able to go backwards.

- An increase in the number of free particles during a reaction helps to drive it forward, because the number of available particle configurations thereby is increased.

3.5   It gets cold
Endothermic reaction

We have seen two examples of spontaneous exothermic reactions and now we shall examine a spontaneous endothermic ("heat absorbing") reaction. It was believed a long time ago that it is the release of energy that is the driving force for chemical reactions, and then the existence of spontaneous endothermic reactions was a mystery. We have already seen examples of spontaneous endothermic processes – one such is the evaporation of a drop of liquid before the saturation pressure is reached. There the driving force is the release of molecules into the gas phase and the resulting increase in the number of possible particle configurations. Here, we shall see what can be the driving force for a spontaneous endothermic reaction.

Barium hydroxide and ammonium nitrate, both of which are solids, react with each other and form solid barium nitrate, liquid water and ammonia:

$$Ba(OH)_2(s)+2\ NH_4NO_3(s) \rightarrow Ba(NO_3)_2(s)+2\ H_2O(l)+2\ NH_3(solution),$$

where (l) stands for "liquid." The ammonia formed becomes dissolved in the water. The reactants have lower energy than the products so energy is absorbed during the reaction (Figure 3.11). This energy is taken from the surroundings which get colder. If one performs the reaction in a vessel immersed in a small amount of water, the water will freeze to ice.

Because the products absorb energy[12] from the surroundings, there is a reduction in the number of available microstates with different energy distributions. This leads to a negative contribution to the total entropy, which is unfavorable. Since the process is spontaneous,

---

[12]The energy is stored in the products as potential energy in the interaction between the electrons and the nuclei and as kinetic energy of these particles.

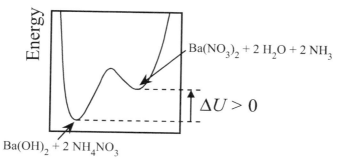

Figure 3.11   Schematic energy diagram for the reaction between the solids barium hydroxide and ammonium nitrate. The products are solid barium nitrate and an aqueous solution of ammonia. The energy difference ΔU between the products and reactants is positive, so energy is taken from the surroundings during the reaction.

there must be a positive contribution to the entropy that dominates over this negative contribution. From what does it arise?

Both reactants are solid salts whose ions are located in crystal lattices. The number of available particle configurations for them is low, so these substances have a relatively low entropy. This also applies to the product barium nitrate, which is also a solid salt. However, an aqueous solution of ammonia is formed during the reaction and this has a high entropy because its molecules are free to adopt a wide variety of particle configurations. The end result is that the total entropy increases during the reaction.

Key points

- An endothermic reaction can occur spontaneously provided the loss in entropy, which is a consequence of the absorption of energy from the environment, is compensated by some larger gain in entropy, for example by particles being released into solution or into a gas phase during the reaction.

3.6 Colloidal stability
Repulsion driven by entropy

A colloidal particle is a particle with a size on the order of 1 to 1000 nm.[13] Such particles can often be dispersed in a liquid and remain suspended in the liquid without falling to the bottom or floating up to the surface. Examples of colloidal dispersions from everyday life are milk, mayonnaise, and mud (like wet clay). If the colloidal particles clumped together and formed large aggregates, the dispersion would not be stable. When the aggregates became large enough, they would fall to the bottom if their density was greater than that of the liquid or float to the surface if their density was lower. This happens, for example, when milk is curdled to make cheese. The colloidal particles of milk, which consist mainly of fat and protein (casein), clump together when one adds some curdling agent (usually rennet) and one obtains large aggregates, which form curd. One can also make milk curdle by for example adding an acid.

In order for a dispersion to be stable, there must be something that prevents the colloidal particles from being lumped together – a repulsive force between the particles that prevents them from forming aggregates. A common reason for such a repulsive force between the particles is that they become electrically charged when dispersed in water, and we shall discuss a simple example of such a case.

The particles, which initially are electrically neutral, may become charged when dispersed. This may happen because of ionization of molecular groups on the surfaces of the particles. One example is carboxylic acid groups that release their protons into the solution, whereby the negatively charged carboxylate groups remain on the particle, which hence becomes negatively charged. In other cases, ions that were bound to the surface by ionic bonds can be dissolved by the water, whereby the oppositely charged groups on the surface remain, like Figure 3.12. Yet another reason for a particle to become charged can be that some ions from the aqueous phase become adsorbed on the particle surface. If a larger number of negatively than positively charged ions are adsorbed, the particle will obtain a negative net charge.

Let us assume that the particles have positive ions adsorbed at the surfaces before they are dispersed in water and that negative molecular groups are firmly attached to the surfaces. For simplicity, we draw

---

[13] 1 nm = 10^{-9} m.

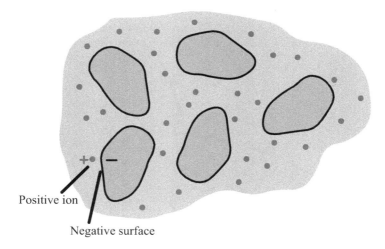

Positive ion

Negative surface

Figure 3.12   Example of colloidal dispersion where the surfaces of the colloidal particles are negatively charged and where positive ions exist in the surrounding aqueous solution.

in Figure 3.13 two adjacent particles as if their surfaces were flat and parallel. The colloidal particles are electrically neutral in the dry state (they have the same amount of positive and negative charges). The electrostatic interactions between them are quite weak in this state because the positive and the negative charges sit near each other on each surface.

When the particles are dispersed in water, most ions become detached from the surfaces (Figure 3.14). This is mainly due to the fact that the water molecules hydrate (bind to) the ions and that the electrostatic interaction between the charges is reduced significantly due to the surrounding water (about 80 times weaker interaction in water than in air).

The electrostatic energy would be minimized if all ions were sitting at the surfaces, since the positive and negative charges would then be as close together as possible. The entropy of mixing would, on the other hand, be maximized if the ions would distribute themselves evenly in the gap. The equilibrium state is a compromise between energy minimization and entropy maximization, whereby the distribution of ions becomes something in between these two extremes.[14] The

[14]Just like before, energy minimization means that the released energy is spread over the system and the surroundings (a gain in entropy). The equilibrium state corresponds to the maximal total entropy.

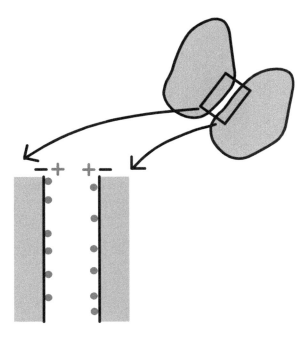

Figure 3.13   Enlarged sketch of the gap between two neighboring particles and parts of the surfaces. The positive ions are bound to the surfaces of the particles when they are in dry condition.

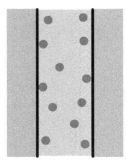

Figure 3.14   The ions detach from the surfaces when the particles are dispersed in water.

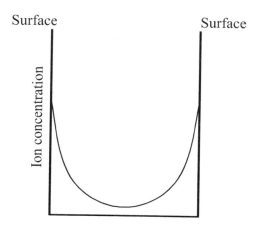

Figure 3.15   A sketch of the ion concentration in the gap between the surfaces of the particles.

released ions will move around in the gap between the particles, but they will mostly be located in the vicinity of the negative surfaces since they are attracted electrostatically to the latter and repelled from each other. The ion concentration is therefore highest at the surfaces and lowest in the middle – a concentration profile is formed in the gap at equilibrium, as shown in Figure 3.15. At each surface there are accordingly a negative surface charge and a "cloud" of positive ions just outside it. This is called a **diffuse electrical double layer** – an important concept in surface and colloid chemistry.

What happens now if we increase the distance between the surfaces? The volume in the gap between the surfaces will increase, whereby the ions will have a larger space to move within. This means that the number of available particle configurations increases, leading to a positive entropy contribution. In most cases of interest, this is the most important effect of the increase in distance. Since the entropy increases, this is advantageous for the system. Therefore, if the colloidal particles are free to move, the distance between them increases spontaneously, that is, a repulsive force acts between the surfaces (Figure 3.16). This force, called an **electric double layer force**, is the reason why many colloidal dispersions are stable. Note that it is this entropy

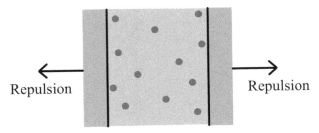

Figure 3.16   Upon an increase of the distance between the surfaces the configurational entropy of the ions increases. This entropy increase manifests itself as a repulsive force between the surfaces.

increase and not an electrostatic force between the colloidal particles that causes the repulsion.[15]

Key points

- A repulsive force between like-charged surfaces in an electrolyte solution arises because the entropy of the ions in the gap between the surfaces increases as the distance between the surfaces is increased and the available volume for the ions thereby becomes larger.

3.7   What is the driving force?
Total entropy of the system and the surroundings

In several very different cases, we have seen that the same principle recurs for the driving force of spontaneous processes. The principle is usually expressed as an increase of the total entropy of our system and the surroundings for all spontaneous processes.

However, we have seen in the various cases, how this is a manifestation of the fact that the macroscopic equilibrium state is determined

---

[15]For planar, equally charged surfaces, each half of the system on either side of the mid-plane is electroneutral (on average). There are as many positive charges (ions) as there are negative charges (on the surfaces). One can show that fluctuations, that temporarily make each half deviate from electroneutrality, lead to an attractive(!) electrostatic contribution to the force between the like-charged surfaces (even fluctuations that maintain electrical neutrality of each half can provide attractive contributions). Often (but not always) this attraction is, however, smaller than the repulsion, so the net force is repulsive.

by the most probable distributions of energy and particles, that is, the microstates that are macroscopically indistinguishable and are overwhelmingly more numerous compared to microstates with other distributions. Since entropy can be expressed in terms of the number of microstates by the Boltzmann relation $S = k_B \ln \Omega$, we obtain the link between the microscopic course of events and the macroscopic quantity entropy.

It must be strongly emphasized that the principle of increasing entropy only applies to the *total entropy*, that is, the entropy of the system *and* the surroundings. The entropy of the system itself may increase or decrease depending on the circumstances. For an *isolated system*, that is, a system that cannot exchange energy or particles with the surroundings, the principle of increasing entropy applies to the system itself (the entropy of the surroundings will not change due to the process in this case).

Since we actually are mostly interested in our system, it would be handy to avoid having to worry about the surroundings. We would then be able to concentrate solely on the system itself. In several of our examples, we saw how we can alternatively express the equilibrium state as a compromise between low energy and high entropy *of the system*, whereby we only consider energy and entropy changes for our system. The correct way to handle this balance between energy and entropy of a system is to introduce the concept of **free energy**. This is what we shall do in the next section.

Ultimately, it is the increase in total entropy that is the main criterion for spontaneous processes, but it is not the increase in entropy itself that is the driving force. The entropy is a collective property of the entire macroscopic system and the individual molecules "do not know of" its existence. The molecules do everything that is possible for them to do under the circumstances: they move around freely provided nothing prevents them from doing so, and they exchange energy back and forth – all in a random manner. The end result is that the most likely happens most often. This is what the entropy increase expresses.

What then is the actual driving force in the world of the molecules? One could say that it is simply "blind" opportunism. The molecules are free to do everything that is possible and they will do so sooner or later! The final result that we observe is the outcome that has an overwhelming probability to occur.

Key points

- The general criterion for a process to be spontaneous is that the *total entropy* of the system and the surroundings increases.

- For an *isolated* system, the criterion for spontaneous processes is that the entropy of the system itself increases. In all other cases, the entropy changes in the surroundings must also be included.

- The increase in entropy is a manifestation of the fact that the macroscopic equilibrium state is determined by the microstates that are macroscopically indistinguishable and are overwhelmingly most numerous. This macrostate is therefore much more probable than other macrostates and has largest total entropy.

- Spontaneous processes go in the direction that is most likely: *the most probable happens most often*. This is what the entropy increase expresses.

3.8   To indirectly keep track of the surroundings
The concept of free energy

If a system is isolated, the criterion for a process to occur spontaneously is, as we have seen, that the entropy of the system increases. In most cases, however, one is in practice interested in systems that are not isolated, such as systems that can exchange energy with the surroundings. In such cases, one must use the general criterion of increase in the total entropy, that is, the sum of the entropy of the system and the surroundings becomes greater when a process occurs spontaneously.

An important case is a system that one holds at a *constant temperature* by allowing it to exchange heat with a **thermostat**, that is, a different system with the property that it maintains its temperature T at a fixed value. If an exothermic process takes place in our system, heat must be removed (from the system to the thermostat) for the system to remain at constant temperature. Correspondingly, if an endothermic process occurs, heat must instead be supplied by the thermostat. By means of free heat transfer between our system and the thermostat, the temperature of the former is automatically kept equal to that of the latter (Figure 3.17). Usually the surroundings of the system are

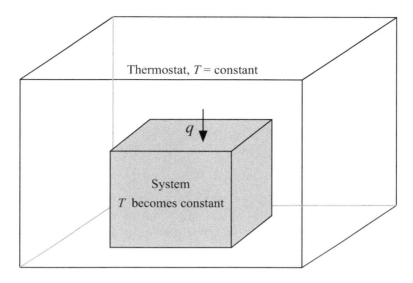

Figure 3.17   A system in contact with a thermostat obtains constant temperature when heat is freely exchanged between the system and the thermostat.

considered to be a thermostat provided there is a constant temperature. If the surroundings are large enough, the heat exchange with the system does not affect the temperature appreciably, so T is therefore the same before and after the process. When the system is free to exchange heat with the surroundings, the system's initial and final temperature is also T.

Let us assume that the entropy of the system changes by the amount ΔS during the process and that an amount of heat q must be supplied *to* the system in order to maintain its temperature T. If the process is endothermic $q > 0$ and if it is exothermic $q < 0$ (remember that q is always counted as the heat added to the system, so it is negative when heat is removed). For simplicity, we consider an endothermic process in the following argument, but the end results apply regardless of the sign of q.

When the process takes place in our system, the surroundings accordingly deliver the amount of heat q to the system. Since there remains less energy to distribute over molecules in the surroundings, its entropy has decreased. Specifically, the entropy of the surroundings is reduced by the amount q/T; we have $\Delta S_{surr} = -q/T$. This follows from Equation (2.15) applied to the surroundings: $\Delta S_{surr} = q_{surr}/T$, where

$q_{surr} = -q$ is the amount of heat that the surroundings *receive* during the process (a negative amount in our case). The total entropy is therefore changed by

$$\Delta S_{tot} = \Delta S + \Delta S_{surr} = \Delta S - \frac{q}{T}. \tag{3.1}$$

The process is spontaneous if $\Delta S_{tot} > 0$, that is, if

$$\Delta S - \frac{q}{T} > 0 \quad \text{(for spontaneous process at constant } T\text{).} \tag{3.2}$$

The entropy of the system then increases more than the entropy of the surroundings is reduced (for an endothermic process). This happens, for example, when a drop of water in a small closed box evaporates at constant temperature (see the previous example in Section 3.2). The increased configurational entropy when water molecules leave the drop and vaporize provides a large positive contribution to ΔS, while the energy used to release molecules from the drop is taken from the surroundings of the box whereby $\Delta S_{surr} = -q/T < 0$. The water evaporates spontaneously when the reduction in the entropy of the surroundings is less than the increase of the entropy in the box.

Equation (3.2) is also valid when q is negative and ΔS_{surr} therefore is positive, for example, an exothermic combustion process in a closed box that can exchange heat with the surroundings. The most favorable case occurs when both ΔS and ΔS_{surr} in Equation (3.1) are positive, whereby the entropy changes in the system and the changes in the surroundings work together and provide positive contributions to the total entropy change. One such case is the combustion of stearic acid in oxygen (as in Section 3.4 but performed in a box at constant T, i.e., with the same initial and final T). In this case, the entropy increase of the system acts together with the release of heat to the surroundings, which also gives an entropy increase. Both effects contribute to making the process spontaneous.

The condition (3.2) can be satisfied even if the entropy of the system decreases ($\Delta S < 0$), provided that the system releases such a large amount of heat (q is strongly negative) that the entropy increases more in the surroundings than it decreases in the system. In such cases, for example, combustion of magnesium in oxygen (Section 3.3), it is the release of heat to the surroundings that makes the process occur spontaneously. The decrease in entropy of the system is to a large part due to the binding of oxygen molecules to magnesium, whereby the number of configurations is reduced.

The great advantage of the condition (3.2) is that all variables involved describe characteristics of the system itself: q is the amount of heat that the system receives during the process, ΔS is its entropy change and T is the initial and final temperature (which is equal to the constant temperature of the surroundings). Equation (3.2) is accordingly a condition that does not involve the surroundings explicitly (except via the temperature), and despite this it is equivalent to the increase in total entropy (the sum of the entropy of the system and the surroundings). This fact can be made even clearer when, for example, the system can exchange energy with the surroundings only in form of heat.[16] Under this condition, the quantity of heat q is equal to the change in energy ΔU of the system during the process (as we shall see later, this assumes that the system volume V is constant). By inserting $q = \Delta U$ in Equation (3.2) we obtain

$$\Delta S - \frac{\Delta U}{T} > 0 \quad \text{(for spontaneous process at constant } T \text{ and } V), \quad (3.3)$$

a condition that only involves quantities for the process in our system (ΔU and ΔS) and the temperature T at which it takes place. This condition implies that $\Delta S - \Delta U/T = \Delta S + \Delta S_{\text{surr}} = \Delta S_{\text{tot}} > 0$ for a spontaneous process when T and V are constant.

We can bring these arguments one step further. Equation (3.1) with $q = \Delta U$ can be written

$$\Delta S_{\text{tot}} = \Delta S - \frac{\Delta U}{T} = \frac{T\Delta S - \Delta U}{T} = -\frac{\Delta U - T\Delta S}{T}, \quad (3.4)$$

which means that the *total* entropy change can be expressed in terms of the energy and entropy changes of the system. Since T is the same before and after the process we have

$$
\begin{aligned}
\Delta U - T\Delta S &= (U_{\text{after}} - U_{\text{before}}) - T(S_{\text{after}} - S_{\text{before}}) \\
&= U_{\text{after}} - U_{\text{before}} - T S_{\text{after}} + T S_{\text{before}} \\
&= (U_{\text{after}} - T S_{\text{after}}) - (U_{\text{before}} - T S_{\text{before}}) \\
&= (U - TS)_{\text{after}} - (U - TS)_{\text{before}}
\end{aligned}
$$

and we can write Equation (3.4) as

$$\Delta S_{\text{tot}} = -\frac{(U - TS)_{\text{after}} - (U - TS)_{\text{before}}}{T}, \quad (3.5)$$

---

[16]Another example is given in Section 4.7, where the corresponding discussion is given for constant pressure – a more common case.

at constant T and V. This equation says that the total entropy change is proportional to the change in the value of $U - TS$, that is, the number one obtains when subtracting the product of the system's temperature and entropy from its energy. If the value of $U - TS$ decreases during the process, ΔS_{tot} is positive, and if $U - TS$ increases ΔS_{tot} is negative. We can therefore determine whether a process can occur spontaneously or not by examining the change in $U - TS$. This combination of variables has accordingly an important role to play and it is therefore convenient to introduce a separate symbol A for it,

$$A = U - TS, \tag{3.6}$$

and give it a name: "free energy" (more specifically, the **Helmholtz free energy** or simply the **Helmholtz energy**).[17] Thus we have from Equation (3.5)

$$\Delta A = -T\Delta S_{tot} \quad \text{(at constant } T \text{ and } V). \tag{3.7}$$

As we have seen, the crucial factor that determines whether a process in the system is spontaneous or not, is the balance between the entropy change ΔS and the effect of the energy change ΔU of the system (the latter in terms of change of the entropy of the surroundings). The free energy A is so constructed that it "automatically" handles this balance for us,

$$\Delta A = \Delta U - T\Delta S = -T\Delta S_{tot} \quad \text{(at constant } T \text{ and } V), \tag{3.8}$$

so if $\Delta A < 0$ the process *can* take place spontaneously ($\Delta S_{tot} > 0$), and if $\Delta A > 0$ the process *cannot* take place spontaneously ($\Delta S_{tot} < 0$). In the latter case, the reverse process can instead occur spontaneously since it has the opposite sign of ΔS_{tot}.

The principle behind Equation (3.8) is so important that it is worth repeating. The change in free energy ΔA constitutes a quantity that correctly balances the changes in energy ΔU and entropy ΔS of the system. Thereby the contribution ΔU describes the effect of the change in the entropy of the surroundings, $\Delta U = -T\Delta S_{surr}$. Accordingly, $\Delta A = \Delta U - T\Delta S = -T\Delta S_{surr} - T\Delta S_{system} = -T\Delta S_{tot}$. Hence, the

---

[17]The modern, recommended name is the Helmholtz energy, but Helmholtz free energy, which has long been the common name, is such an established concept that it is not eradicated easily. Furthermore, it is convenient to speak of "free energy" to distinguish from "energy" which does not take entropy into account.

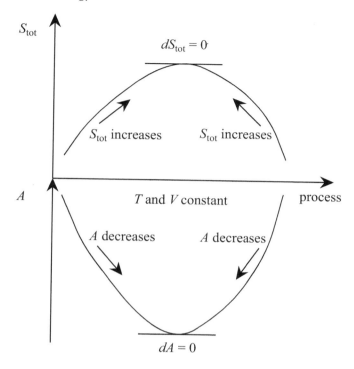

Figure 3.18 A process at constant temperature and volume can occur spontaneously in the direction of decreasing free energy A (Helmholtz energy). Thereby, the total entropy S_{tot} of the system and the surroundings increases. Equilibrium is reached when A has reached its minimal value and at the same time S_{tot} is as large as possible under the given circumstances. At the minimum and the maximum point, respectively, the derivative is zero (horizontal line) and $dA = 0$ and $dS_{tot} = 0$, respectively.

Helmholtz energy keeps track of the entropy changes of both the system and the surroundings when T and V are constant.

If we let the process take place in small steps, it will occur spontaneously as long as the total entropy increases; that is, as long as the free energy decreases (see Equation (3.7)). Because the change occurs in small steps, we denote the change in the total entropy by dS_{tot} and the free energy by dA (small changes, called differentials). Thus, we have $dA = -TdS_{tot}$, where $dS_{tot} > 0$ and $dA < 0$.

Eventually, the process will halt. This occurs when the total entropy has become as large as possible. It has reached its maximum.

Simultaneously, the free energy of the system has decreased as much as possible; it has reached its minimum, as illustrated in Figure 3.18.

The system has thus attained the most likely macroscopic distribution of energy and particles under the prevailing circumstances (circumstances = "the system is in a closed box and has constant volume and temperature"). For this distribution, there are tremendously more microstates of the system and the surroundings than for all other distributions, and hence this macroscopic state has maximal total entropy under the given circumstances. From the preceding discussion it follows that the *system* simultaneously has attained its minimum free energy. The state that is reached is the macroscopic equilibrium state of the system.

Assume now that the system is initially in its equilibrium state. If the process were to go in any direction, S_{tot} would decrease because the point of departure is at the maximum value. Simultaneously, A would increase from its minimum value. Such a change cannot occur spontaneously because spontaneous processes require that S_{tot} increases (A decreases). The only way for us to make the system leave the equilibrium state is by forcing a change by taking action from the outside and altering, for example, the amount of molecules of some kind.[18] This change cannot occur spontaneously when the system is left alone.

If, instead, we try to make the process take a *tiny* fraction of a step from the equilibrium state, the total entropy and the free energy will not change significantly, $dA = 0$ and $dS_{tot} = 0$. This is characteristic for the equilibrium state and we take

$$dA = 0 \quad \text{(equilibrium at constant } T \text{ and } V) \tag{3.9}$$

as a condition of equilibrium (this corresponds to the horizontal tangent in Figure 3.18). Tiny deviations from the equilibrium state (such as small spontaneous fluctuations) are thus allowed and happen in reality. Typically, they are incredibly small for a macroscopic system.

The concept of free energy is one of the most important ones in thermodynamics and we shall return to it later. The Helmholtz energy A, which we have discussed here, has a central role in determining the direction of spontaneous processes and in specifying the

---

[18] In the example of magnesium this can be done for instance by decomposing MgO into O_2 molecules and Mg, which requires that energy is actively supplied from the outside.

condition for equilibrium of systems at constant temperature and volume. However, it is not often that one is interested in systems with constant volume. More common are processes that are carried out, for example, at constant pressure whereby the volume is allowed to vary. It is therefore of interest to examine what happens when one changes the volume of a system. We shall do this in the next chapter. This is particularly important when we have a gas phase present because the volume of gases changes a lot when the number of gas molecules is changed, such as when a chemical reaction occurs that involves gaseous substances.

Key points

- Free energy is a system property which is designed so that it keeps track of the entropy changes in both the system and the surroundings (provided that certain parameters, such as temperature, are kept constant).

- The free energy **Helmholtz energy**, A, is defined as

$$A = U - TS.$$

- When T and V are constant we have $\Delta A = -T\Delta S_{\text{tot}}$.

- The criterion for a process to be spontaneous at constant T and V is that A decreases, $\Delta A < 0$.

- Equilibrium at constant T and V occurs when A has reached its minimal value. Then S_{tot} is as large as possible. At the minimum point for A we have $dA = 0$.

CHAPTER 4

More on gases and the basics of thermodynamics

4.1 Bike pumps and fridges
Gas compression, pressure, and work

Anyone who has pumped up a bicycle tire by hand knows that the bike pump becomes hot. Part of the increase in temperature is due to friction, but there is also another reason for the increase. In fact, a gas heats up when it is compressed. By depressing the bike pump handle and thereby pushing the piston inwards, one compresses the gas in the pump and the gas pressure increases. When the pressure becomes sufficiently large, the gas enters through the valve on the bike tube and the pressure in the tube increases – which is the purpose for pumping. The temperature increase is a side effect that one does not really want. However, it is an interesting phenomenon which has been put to use in other everyday contexts. One example is a diesel engine, where air that has been drawn into a cylinder is compressed strongly when the piston moves inwards. Thereby the temperature of the air increases so much that when oil is injected into the cylinder, it ignites spontaneously. The resulting combustion gases pushes the piston out again. The alternating motion inwards and outwards of the piston drives the engine.

Before investigating the cause for the temperature increase during compression, let us first look at some more examples of practical applications of this phenomenon. In a common type of refrigerator (so-called compression refrigerator), shown as a schematic in Figure 4.1, the cooling takes place when a liquid (the refrigerant) evaporates at low pressure (A) in an evaporator that is placed inside the refrigerator. For molecules to be able to leave the liquid and vaporize, energy is required as we have seen in Section 3.2. This energy is brought in the form of heat from inside the fridge, which thereby is cooled down.

The gas that is formed in the evaporator is drawn (B) into a compressor which compresses the gas strongly (C). Thereby, the temperature of the refrigerant gas increases so that it becomes much higher

83

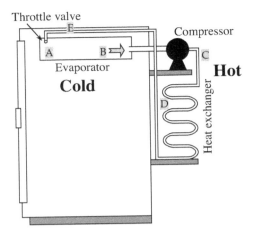

Figure 4.1   A schematic diagram of a refrigerator. Liquid refrigerant passes out through a narrow valve (throttle valve) and vaporizes (A). The heat required for vaporization is taken from the refrigerator's interior. The steam enters (B) into a compressor (C) where its temperature is raised greatly during compression. The hot steam is cooled (D) in a heat exchanger outside the fridge and is condensed into liquid that is returned (E) to the throttle valve.

than room temperature. This is accordingly a practical example of a rise in temperature during compression. The hot gas is led through tubes that are cooled by the surrounding air (a heat exchanger) and heat is dissipated from the compressed gas (D) into the room where the fridge is situated. Thus we obtain refrigerant gas at high pressure but at a considerably lower temperature. As we saw in Section 3.2, the gas will condense when the vapor pressure of the liquid is less than the pressure of the gas (just like when water droplets are formed, for example, in the form of fog when moist air is cooled). This happens when the temperature becomes sufficiently low. During the condensation a large amount of heat is released, which is also dissipated into the room through the heat exchanger. The liquid is then led (E) through a narrow valve (throttle valve) to the evaporator, where it vaporizes at low pressure and the process is repeated. The temperature rise during gas compression therefore has a key role in the refrigeration process, even if the main transfer of heat is via first evaporation and then condensation of the refrigerant.

Temperature increase due to gas compression is also essential in heat pumps. Such a pump is basically a reverse refrigeration unit

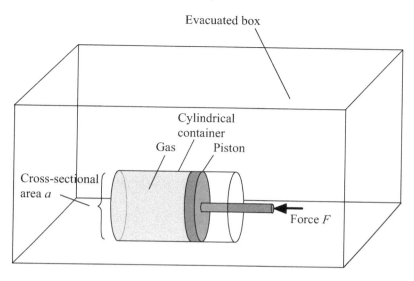

Figure 4.2 A cylinder that is sealed by a freely movable piston, which is assumed to run without any friction. The cross-sectional area of the cylinder and the piston is *a*. The device is located in an evacuated box. The cylinder contains a gas. The volume of the gas can be changed by moving the piston inwards or outwards. The piston is held at the desired position by the application of a suitably large force *F* on the exterior side of the piston. When *F* is increased the gas is compressed and when *F* is reduced it expands.

where the heating occurs, for example, inside a house and the cooling outside. In this case, it is the temperature increase that one wants for heating the building. Here one uses energy to pump heat from the cold surroundings to the warm(er) house. The heat released in the building comes both from the surroundings outside (which is cooled) and from the electrical energy used to drive the pump. Therefore, less electrical energy is used to bring a certain amount of heat to the inside, than if the same amount of heat would be generated directly in an electric heater. This is the reason for using a heat pump instead of a heater.

Why is the temperature increased when a gas is compressed? Let us consider a cylindrical container with a movable piston where the container volume can be changed by moving the piston, as shown in Figure 4.2. In the cylinder there is a gas. We assume that the piston can move without any friction in the cylinder. The piston and the interior of the cylinder both have a cross-sectional area *a*. For simplicity, we

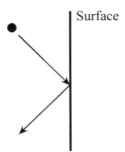

Figure 4.3   A molecule colliding with the surface of a much heavier body changes the direction of its velocity.

assume that the entire device is located in an evacuated box. We apply a force F on the piston.

The gas molecules collide with the container walls and the inside surface of the piston. Each molecule that collides gives rise to a force on the surface since the molecule changes direction upon impact, as illustrated in Figure 4.3. As mentioned in Section 2.1, each individual molecular collision gives a very small contribution to the force on the surface. The total force due to collisions will, however, be very large since there is a huge number of molecules colliding. What we can measure macroscopically is an average value of the total force, which does not vary noticeably in time but is constant.

The force that acts on the surface due to molecular collisions is proportional to the surface area. A surface area twice as large is exposed to force twice as large since the number of molecular collisions is doubled (given that the gas density is equal in the two cases). The force on the surface from the gas molecules divided by the area of the surface is, however, the same in both cases and it is this quantity that is called the pressure, P, of the gas

$$P = \frac{\text{Total force on surface}}{\text{Surface area}}.$$

Thus, in our container above, the interior surface of the piston will be exposed to a force due to the gas that is equal to $a \times P$. This force is directed to the right (perpendicularly to the piston surface).

To obtain a feeling for the order of magnitude, one can easily figure out that a gas with a pressure of 1 atmosphere (1 atm $\approx$ $101300\ \mathrm{Nm^{-2}}$) exposes an area of 3 dm^2 for a force of 3040 N, i.e., the

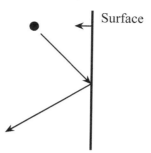

Figure 4.4 The speed of a molecule increases when it collides with a body that is moving towards the molecule. This occurs during compression of a gas when the piston moves inwards.

gravitational force on a 310 kg weight at Earth's surface. Since 3 dm^2 is approximately the area of the top surface of the head, it is as if we humans go around carrying 310 kg on top of our heads because of the atmosphere. We do not experience this great weight because our tissues have an internal pressure that exactly counteracts the external pressure.

Similarly, we have to apply a force $F = aP$ on the outside of the piston in Figure 4.2 for it not to move, that is, the force on the outside and inside of the piston must be equal. As long as the force F has this magnitude, nothing will happen (provided everything else is unchanged). If we want to compress the gas, we have to exert a force $F > aP$ and if we want to expand the gas, we need to reduce the force so that $F < aP$. The piston will then move and the pressure of the gas will change until the forces on either side of the piston are equal again.

What happens when we exert a force $F > aP$ so that the piston moves to the left and the gas is compressed? For simplicity, we limit ourselves to monatomic molecules such as argon, but the principles are largely the same for polyatomic molecules. We also assume that the molecular collisions with the piston are elastic, which means that they occur without changes in kinetic energy (other cases are treated later). Since the piston moves to the left, each molecule that collides with the piston surface will acquire a higher speed after the collision compared to before (Figure 4.4), just like a tennis ball receives a higher speed when we hit it with a tennis racket (see Appendix C for a detailed discussion about this). The influence of each collision on the piston's speed is, on the other hand, extremely small, since the piston has a vastly larger mass than the molecule.

The gas molecules move generally much faster than the piston, so their increase in speed is relatively small at each collision, but each molecule will collide several times with the piston surface and after several collisions, its speed will have increased significantly. Since the molecules also collide with each other, the changes in speed will spread among all molecules.[1]

The movement of the piston to the left thus results in an increase in speeds of the gas molecules and therefore an increase in their kinetic energy. It is this energy increase that causes the increase in temperature. If the piston and the cylindrical container are made of thermally insulating materials, the energy will stay in the gas and the temperature will remain high. Otherwise, the energy will spread to the surroundings by, for example, heat conduction through the walls of the container and of the evacuated box.[2] This happens until the gas temperature has again been reduced to that of the surroundings.

How much energy have we added to the gas during compression? Fortunately, we do not need to keep track of how much energy each molecular collision transfers; it is enough to keep track of the force F (which we use to compress the gas) and the distance the piston moves. If we push on an object with force F and the object moves a distance s, we have, according to the laws of mechanics, done the **work** $w = F \times s$, which is equal to the energy that we have added to the object. Because it is we ourselves who control the force F that acts on the piston and since we know how far the piston is moved, we also know how much work we have performed.

We assume that the gas is thermally insulated from the surroundings, so the only energy exchange it has is the work w that we add during the compression. Let us do the compression with a constant force F (Figure 4.5). If the piston thereby moves the distance s, the

---

[1]For polyatomic molecules one must also take into account that they rotate and vibrate, while monatomic molecules cannot do so. A monatomic molecule's energy is essentially equal to its kinetic (translational) energy, which changes when its speed is changed. (We here ignore the energy due to the atom's internal structure, which does not change in the present context. If the temperature is moderately high or less, one can in most cases ignore the energy that is received through electronic excitations.)

[2]The energy can reach the walls of the box from the cylindrical container via radiation even when the box is completely evacuated.

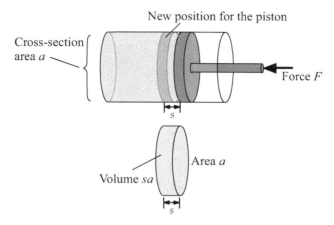

Figure 4.5   During a compression where the piston moves distance s to the left, the gas volume decreases by the amount sa, where a is the cross-section area of the cylinder. The size of this volume is illustrated below the cylinder.

energy of the gas has been changed by[3]

$$\Delta U = w = Fs. \tag{4.1}$$

Note that even if the piston moved very slowly, the energy of the gas will change by the amount w according to Equation (4.1). The molecular speeds will then increase by a very small amount for each collision, but it will take a long time before the piston has moved the distance s so the molecules will have time to collide a very large number of times.[4]

Since the cross-section area of the cylinder is a, the gas volume is reduced by the amount $s \times a$ during compression, as illustrated in Figure 4.5. The gas volume has decreased from V_{before} to $V_{after} = V_{before} - sa$, which means that $\Delta V = V_{after} - V_{before} = -sa$. Note that ΔV is negative since we decrease the volume. It follows that $s = -\Delta V/a$, so we have

$$w = Fs = -\frac{F\Delta V}{a}. \tag{4.2}$$

---

[3]In this book, the work w constitutes an energy added *to the system*. There exists another common convention for w as explained in the footnote to Equation (4.6).

[4]No matter how fast the piston moves, we assume that all kinetic energy that the piston receives through the action of the force F is transmitted to the gas.

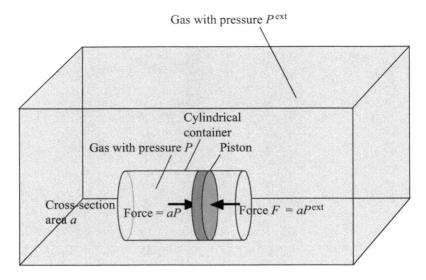

Figure 4.6   When the cylinder with the movable piston is placed in a box filled with gas, a force $F = aP^{ext}$ acts on the outside of the piston, where P^{ext} is the gas pressure in the box. The gas inside the cylinder acts with the force aP on the inside of the piston, where P is the gas pressure in the cylinder. When these two forces are equal, there is equilibrium and the piston is not moving. Otherwise the piston will be set in motion in the same direction as the greatest force.

We add energy in the form of work to the gas during compression and, consequently, the work w is positive when $\Delta V < 0$ (therefore there is a minus sign in the formula).

In our example we have applied the force F via a rod attached to the piston. This is obviously not necessary. We can, for example, let the cylinder be in a gas-filled box with the gas pressure P^{ext} ("ext" stands for external), whereby the gas exerts a force $F = aP^{ext}$ on the outside of the piston, as shown in Figure 4.6. If the box is open, P^{ext} is equal to the ambient pressure of the atmosphere.

Just like before, the piston will not move if the force on either side of the piston is equal, that is $F = aP$, where P is gas pressure inside the cylinder. Thus, we have the condition $P = P^{ext}$ for this to apply, i.e., the gas pressure is equal on both sides at equilibrium. If $P < P^{ext}$ the gas in the cylinder will compress and if $P > P^{ext}$ it will expand. This will occur until the pressure is the same on the inside and the outside.

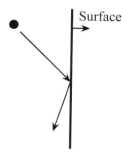

Figure 4.7 The speed of a molecule decreases when it collides with a body that moves in the same direction as the approaching molecule. This occurs during expansion of a gas when the piston moves outwards.

If we compress the gas at a constant external pressure, P^{ext}, such as the atmospheric pressure, we obtain from Equation (4.2) with $F = aP^{ext}$ inserted

$$w = -P^{ext}\Delta V. \tag{4.3}$$

Equation (4.3) also applies during expansion (when $P^{ext} < P$). In such cases, the gas volume in the cylinder will increase, that is, $\Delta V > 0$, and the work w is negative. When the gas expands, it performs work on the surroundings, which is the same thing as the surroundings performing negative work w on the gas. The energy of the gas in the cylinder is thereby decreased, $\Delta U = w < 0$.

This decrease in energy occurs because when the piston moves to the right, the molecules will acquire *lower* speeds at collisions with the piston surface and thereby lower kinetic energy (Figure 4.7). (Compare this with the corresponding situation during compression, Figure 4.4, where the wall motion to the left gave increased speed of the molecules. For further discussion, see Appendix C.)

Molecules that collide with the inner surface of the piston will set it in motion to the right. This continues as long as the force aP on the inside of the piston is greater than the opposing force $F = aP^{ext}$ on the outside. Part of the kinetic energy of the gas in the cylinder is thus used to move the piston against the external force F, and this energy is accordingly transferred to the surroundings. The gas pressure in the cylinder is reduced during the expansion, and at equilibrium it has become equal to the external pressure.

Finally, we will look at cases where the volume changes by small amounts. For a small change in volume, dV, the work is

$$dw = -P^{\text{ext}}dV, \tag{4.4}$$

whereby the amount of energy $dU = dw = -P^{\text{ext}}dV$ is added to the system (i.e., the gas in the cylinder). The expression (4.3) for the work is valid *only* when the external pressure is constant throughout the entire volume change from V_{before} to V_{after} (with $\Delta V = V_{\text{after}} - V_{\text{before}}$), while Equation (4.4) is general and can be used when the pressure varies. In the latter case, one makes a sequence of small volume changes from V_{before} to V_{after} and sums up the work dw from all of them to obtain the total work. Thereby one uses the appropriate value of P^{ext} for each step.[5] Work performed by a volume change as expressed by Equations (4.3) and (4.4) is called **pressure and volume work** (PV work).

Key points

- The gas pressure P is equal to the force per unit area that the gas molecules exert on a wall surface when they collide with it.

- When a gas is compressed through the action of an external force, such as via a piston, the gas is supplied energy in the form of work. The external force sets the piston in motion and this motion increases the energy of the gas molecules when they collide with the piston.

- When a gas expands against an external force, such as via a piston, the gas gives away energy in the form of work. This is done when

---

[5]Mathematically, this means that an integration is performed

$$w = \int_0^w dw = \int_{\text{before}}^{\text{after}} dw = -\int_{V_{\text{before}}}^{V_{\text{after}}} P^{\text{ext}}(V)dV,$$

where it is taken into account that the external pressure may vary when the volume changes, $P^{\text{ext}} = P^{\text{ext}}(V)$. An important case is when a volume change is done by varying P^{ext} in the setup shown in Figure 4.6 and equilibrium is maintained so that $P^{\text{ext}} = P$ throughout the whole process (since P depends on the volume, P^{ext} must be varied too). This is an example of *reversible work* – a subject that is treated in Section 4.3. If, on the other hand, P^{ext} is constant during the volume change we have $w = -\int_{V_{\text{before}}}^{V_{\text{after}}} P^{\text{ext}}dV = -P^{\text{ext}}\int_{V_{\text{before}}}^{V_{\text{after}}} dV = -P^{\text{ext}}\Delta V$, and Equation (4.3) is recovered.

the gas molecules deliver energy to the surroundings via the piston, which is set in motion from collisions by the molecules that thereby lose kinetic energy.

- The work dw done on the gas during a small volume change, dV, is given by $dw = -P^{ext}dV$, where P^{ext} is the external pressure acting on the piston. This kind of work, which is done when the gas volume is changed, is called **pressure and volume work** (PV work).

- When the volume is changed by ΔV and P^{ext} is constant throughout the entire volume change, the work equals $w = -P^{ext}\Delta V$.

- At equilibrium, the pressure P of the gas is equal to the external pressure, $P = P^{ext}$.

4.2   To work and to heat
Definition of work and heat; the first law of thermodynamics

We have seen how one can change the energy of a system in two different ways: by performing work (w) on the system or adding heat (q) to the system. The signs of q and w are always determined from the system's point of view – positive means addition and negative means removal of energy in the form of work or heat. We have shown that one can raise the temperature of a gas by, for example, performing work on it (by compressing the gas) or simply by heating it, that is, to bring it into contact with something that has a higher temperature. In both cases, we increase the energy of the gas molecules – the difference is the way we do it. Upon compression, energy is transferred to gas molecules when some part of the gas container walls (for example, a piston) moves. During heat transfer, energy is also brought from the wall to the gas molecules, but the wall is stationary. Another way to transfer energy in the form of work to a system is, for instance, to stir with a propeller in a gas or liquid. When the propeller surface hits the molecules, they get a higher energy – an energy that is then spread in the system.

Let us examine what happens when a monatomic gas is heated by heat transfer through a container wall that is, for instance, made of metal. When a gas particle collides with the wall it may lose or gain energy. If the wall is hot and its atoms vibrate strongly, it is very likely that an approaching particle will collide with one or a few vibrating wall atoms in such a way that the particle shoots away from the wall

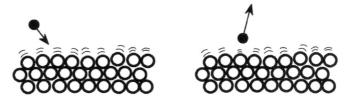

Figure 4.8   A gas particle that collides with a wall consisting of atoms that vibrate strongly (the wall is hotter than the gas) has a high probability to be shot away from the wall at a higher speed than it had before.

at a higher speed, and thus higher kinetic energy, than it had when it approached (see Figure 4.8). As a result, the wall atoms that the particle has hit will vibrate less strongly – they have lost energy to the gas particle – but energy from nearby wall atoms will at the next moment be redistributed, which means that the atoms very soon vibrate "normally" again.

If the gas particle has a very high speed when it approaches – a few particles will have this even if the gas is cold – it is likely that it loses energy when it hits the surface and then the wall atoms instead receive energy and vibrate faster. (This is like hitting a string to make it vibrate.) If the gas is colder than the wall, it is, however, more common that the gas particles receive rather than lose energy at collisions with the wall. Thus, net energy is transferred from the hot wall to the cooler gas; the kinetic energy of the gas particles increases and likewise the gas temperature. This is an example of heat transfer to the gas from the environment. (We assume that the wall similarly obtains energy in the form of heat from the rest of the world on the other side of the wall, which in this case is assumed to have a higher temperature than the gas.)

Similarly, if the gas is warmer than the wall, collisions will on average result in a reduction in energy of the gas particles that collide with the surface and, consequently, the wall atoms will vibrate faster. Heat will then be transferred from the gas to the wall and then passed on to the surroundings (q for the gas, which is our system, is negative here). In these two cases, the energy of the gas has increased or decreased, respectively, without any change in the gas volume (the gas container walls are stationary) and we have

$$\Delta U = q \qquad \text{(when } V \text{ is constant)}, \tag{4.5}$$

where q is the total amount of energy that has been added to the gas via the vibrations of the wall atoms. Note that "removal" = "negative addition," so when heat has been removed from the system q is negative and $\Delta U = q < 0$.

If the wall and the gas have the same temperature and the wall is still stationary, no net energy will be transferred by collisions between the gas molecules and the wall atoms (it is thereby equally likely that energy is transferred in one direction as in the other during the collisions). Since no net energy is transferred, $q = 0$.

For polyatomic molecules, energy will be transferred both to and from their vibrational and rotational motions when they collide with the wall. A molecule can, for example, vibrate faster after the collision than before, whereby it has received energy from atoms in the wall, which vibrate somewhat slower until energy has been supplied from the nearby wall atoms. Energy transfer can also occur between the wall and the gas molecules through radiation,[6] whereby vibrational and rotational motions are changed.[7] The same principles regarding transfer of energy at different and same temperatures also apply in these cases.

In the case where the walls are made of a completely heat insulating material, heat is not led through the wall. The system is then fully **thermally insulated**. The outermost surface layer of the inside of the container wall can, however, possibly be heated or cooled by the gas. This requires an insignificant amount of energy that can be completely neglected in the current context and the heat is not passed on through the wall. The surface layer thereby attains the same temperature as the gas without any significant change in energy of the gas and $q = 0$ in practice.[8] Even if a gas molecule changes its energy at each collision with the wall, the effects of the velocity changes will

---

[6]Everyday examples of heating through radiation are to heat food on an infrared cooktop or in a microwave oven. In the first case, the radiation mainly excites vibrational motions of molecules in the food and in the latter rotational motions of water molecules, from which energy spreads to the other molecules in the food.

[7]The electronic states of the molecules are involved only if the frequency of the radiation is high enough (visible light and UV). Electronic transitions can be important if the wall temperature is very high, such as when the material is white-hot. The significant energy transfer occurs usually at the lower frequencies (infrared and microwave), which involves transitions for rotational and vibrational states of the molecules.

[8]A fully thermally insulating wall is obviously an idealization. In reality, there is always some, albeit small, transfer of heat through the wall.

on average be the same as during the elastic collisions we considered in the preceding section.

When the system is fully thermally insulated, energy can be added or removed only in the form of work, for example, by moving a wall (piston) so that the volume is changed. When the walls are *not* thermally insulating, the energy of the system can be changed both in the form of heat (like heat conduction through the walls) and work (for instance, a volume change). Both forms of energy change may occur simultaneously.

What, then, is the essential difference between transferring heat and performing work (like compression) when one changes the energy of a system? During compression, the energies of the gas molecules are changed due to the *macroscopic* movement of the entire piston – a movement that we can completely control. We can, for instance, change the direction of the piston movement whenever we want. During heat transfer, on the other hand, energies of the gas molecules are changed due to the interaction with the individual wall atoms that have *microscopic* movements (vibrations) – movements that we cannot control in a precise manner. The essential difference between work and heat is thus the way in which energy is transferred to the system; in particular, whether the transfer thereby can be directly controlled macroscopically or not. One can express the difference as follows:

Work is energy transfer due to changes in the external macroscopic variables that define the system and that directly affect the motions of the molecules – variables that we can control macroscopically (like volume).[9] **Heat** is *all* other forms of energy transfer, i.e., any energy transfer that is *not* work. The sum of heat and work is thus always

---

[9]A more accurate way to define work is that it is the energy transfer due to changes in the variables that define the system and that affect the equations of motion of the molecules (i.e., the equations themselves or their boundary conditions). Examples of work are energy changes due to an alteration in location of the system's external enclosing surfaces (walls), a change of the volume of or the number particles in the system. If the system is subjected to an external electric field, the energy change due to variations in the field is also work.

equal to the total internal energy change of a system[10]

$$\Delta U = q + w. \tag{4.6}$$

This expression is a common formulation of the **first law of thermo-dynamics**.

In this form, it may look as if the first law would be trivial: the sum of work and what is not work obviously must be everything. However, the first law of thermodynamics expresses another fact that we touched upon in Section 2.1. One of the main principles of physics is that energy cannot be destroyed or created, it can only be transported from one place to another or be converted from one form to another (like from potential to kinetic energy or vice versa). The total energy U of a system can be changed only by adding or removing energy, that is,

The change in energy of a system
= The energy transferred across the system's boundaries,

whereby addition is a positive transfer and removal is a negative transfer. This is the main content of the first law. The splitting of the energy transfer into heat and work in Equation (4.6) is solely a question of a definition of these two concepts.

The difference in internal energy between the initial and final macroscopic equilibrium states of the system, ΔU, is uniquely determined by these states, while this is *not* the case for the division of ΔU into q and w. There are many ways to go between two macroscopic states and ΔU has the same value regardless of the manner in which we do the change. The values of q and w, on the other hand, depend on *how* the change is done. One says that U is a **state function**, while q and w are not.

The concept of a state function generally means a macroscopic physical quantity that is determined by the system's macroscopic state (at equilibrium) and that does not depend on how this state has been reached. Examples of such functions are internal energy U, entropy S, pressure P, temperature T, volume V, number of particles N, and

---

[10]Sometimes this expression is written as $\Delta U = q - w$, where w is the work that the system performs *on the surroundings*. In this book w always means the work done *on the system*. The only difference between the two alternatives is a change in sign of w everywhere.

the number of moles n, as well as quantities that are defined as combinations of these such as the Helmholtz energy $A = U - TS$.[11] When the system changes from one equilibrium state to another, the change in value for each of these quantities is uniquely determined by the system's initial and final states.

Heat and work are accordingly not state functions because they are not determined by the initial and final macroscopic states of the system. They are instead determined by how the process is designed that takes the system between these states.[12] For example, for a liquid one can change the energy by a certain amount by heating it, stirring in it with a propeller, or both. In the first case $q \neq 0$ and $w = 0$, while in the second $q = 0$ and $w \neq 0$, and in the third $q \neq 0$ and $w \neq 0$. However, the sum $\Delta U = q + w$ is equal when the initial and final state, respectively, are identical in the three cases. It is only the division of ΔU into q and w that are different. When energy is added, the thermal motions of the molecules are increased in all three cases and the same applies to other forms of molecular energies (see the discussion in Section 2.1). For a given macroscopic state of the system, the magnitude and intensity of the thermal motions have nothing to do with how the energy was transferred to the system in order to reach this state – irrespectively of whether heat or work was supplied. The molecules "do not remember" how they received the energy.

In thermodynamics, the terms heat and work *only refer to energy transfer*. You cannot say that a system *contains* a certain amount of heat, just as you cannot say that it contains a certain amount of work. Our everyday language is a bit deceptive in this regard; we happily talk about heat as if it were something that exists in a system. In the summer, you may want to avoid the heat outdoors, or during a cold winter, perhaps you like the heat in a sauna. What you then mean is something that is hot, i.e., something that has a high temperature. It is only when you are exposed to the high temperature that you experience heat in the thermodynamic sense, that is, when heat is transferred to your body. "Being hot" is thus not the same as "to contain heat." The latter has no meaning. Nobody would say that an excavator machine contains work just because it can do work. A system

---

[11]Examples of state functions that we introduce later are heat capacity, enthalpy, and Gibbs energy. The latter two are defined as combinations of other state functions.

[12]Heat and work are therefore sometimes called **process functions**. They are also called **path functions** since they depend on which "path" is taken between the initial and final states.

contains simply an amount of energy that can be transferred as heat or work (or both) to another system. How much of the energy transfer that is heat or work depends, as we have seen, on *how* the process is done.

The symbol Δ is used in thermodynamics only to indicate changes in values of state functions, like ΔU, ΔV, and ΔT. Work and heat are denoted w and q without a Δ because they are not state functions.[13] However, one writes dw and dq to denote small quantities of energy that are transferred in the form of work and heat,[14] whereby the first law of thermodynamics is written

$$dU = dq + dw. \qquad (4.7)$$

Key points

• Work (w) and heat (q) constitute different forms of energy *transfers* to or from a system. How much of the energy transfer that is work or heat, respectively, depends on *how* one performs the transfer.

• **Work** is energy transfer due to changes in the external macroscopic variables that define the system and that directly affect the motions of the molecules – variables that we can control macroscopically (like volume).

• **Heat** is all other forms of energy transfer, whereby the sum of heat and work constitutes the entire energy transfer.

• The signs of q and w are determined from the system's point of view (positive for addition and negative for removal of energy).

• The energy of a system can only be changed by energy transfer to or from the system, that is, $\Delta U = q + w$, or for small amounts of energy, $dU = dq + dw$. This is the **first law of thermodynamics**.

---

[13]The erroneous notation Δw and Δq would imply that w and q were properties of the system, which they are not.

[14]To indicate that it is not a question of a small change in a state function such as dU, dV, and dT, one sometimes uses modified symbols $đw$ and $đq$ for small amounts of work and heat.

4.3   To work quickly or slowly
Entropy during volume changes; reversible work and the second law

When we discussed the entropy S for an ideal gas in Sections 2.3 and 2.4 we saw that S depends on the volume. We treated the case when we allowed a gas that was trapped in a small container to expand into a larger space that was initially empty (vacuum). We made the expansion by removing a partition (Figures 2.11 and 2.12) and we found that the entropy changed by

$$\Delta S_{conf} = N k_B \ln\left(\frac{V_{after}}{V_{before}}\right), \tag{4.8}$$

where N is the number of particles, V_{before} is the gas volume before, and V_{after} the gas volume after the expansion. The change in entropy is, as we have seen, a consequence of the change in the number of particle configurations.

This type of expansion is called "free expansion" since the gas, after our removal of the partition, acquires free access to the larger volume without anything being in its way. In such an expansion no work is performed, $w = 0$. The kinetic energy of the ideal gas does not change. No heat is exchanged with the surroundings, $q = 0$. Thus, the total energy of the gas is unchanged, $\Delta U = 0$. In contrast, the entropy is changed by the amount ΔS_{conf} that is positive.

The gas expansion that we discussed at the end of Section 4.1 is not a free expansion since the gas, in order to expand, has to push a piston against a force F. To move the piston and expand the volume of the gas, the gas molecules will lose kinetic energy: the gas performs work on the surroundings and w is negative. This energy loss implies a negative contribution to the entropy of the gas because there is less energy to distribute among the gas molecules after the expansion. At the same time, the increased volume implies an increase in the number of particle configurations and this gives a positive contribution to the entropy. The latter contribution is given by Equation (4.8) for an ideal gas. If one performs the expansion without heat exchange, $q = 0$ (thermal insulation), these two contributions constitute the entire entropy change. The same applies to compression, but then, the signs of the contributions are reversed (the first positive and the second negative).

Thus there is one positive and one negative contribution to the entropy change of the gas. As we shall see, there is one important case where these contributions will exactly cancel each other so $\Delta S = 0$, namely, when we make the volume change *reversibly* and (still) without heat exchange. A reversible process is, as we have seen, a process that is performed in very small steps (infinitely slowly) when the system is in equilibrium the whole time. The second law, Equation (2.18), says that for an arbitrary process that is carried out in small steps, the change in S for each step is

$$dS = \frac{dq}{T} + dS_{\text{irrev}},$$

where dq is the heat added during the step and dS_{irrev} is the entropy increase due to irreversible changes. This relationship implies that $dS = dS_{\text{irrev}}$ when $dq = 0$. The total change ΔS for the entire process is the sum of dS for all steps. For an arbitrary process with no heat exchange ($q = 0$), we thus have $\Delta S = \Delta S_{\text{irrev}}$. When the process is carried out reversibly $\Delta S_{\text{irrev}} = 0$ and hence $\Delta S = 0$. On the other hand, for an irreversible process we have $\Delta S_{\text{irrev}} > 0$ and the entropy of the system increases. An example of an irreversible process with no heat exchange that is familiar to us is a free expansion of an ideal gas where $\Delta S_{\text{irrev}} = \Delta S_{\text{conf}}$ so that $\Delta S > 0$.

Let us now see how it can be true that the entropy is unchanged during a reversible volume change in absence of any heat exchange. We enclose the gas in a cylindrical container with a movable piston that can move without friction. The cylinder is placed upright in a vacuum chamber and we place a weight on the piston, as illustrated in Figure 4.9. The masses of the piston and the weight are chosen such that the volume of the gas has a certain value V at equilibrium. The gravitational force on the weight and the piston is F (the down arrow in the figure) and at equilibrium the gas pressure is $P = F/a$, so the force on the piston from the inside exactly counteracts F. The system is thus at equilibrium when the volume is V. We assume that the cylinder walls and the piston are thermal insulators, so no heat can be exchanged.

Let us examine what happens if the volume of the gas would change by a very small amount from V to $V + dV$ due to a displacement of the piston by a small distance ds. Thereby the piston carries out a small amount of work $dw = F \times ds$ on the gas and

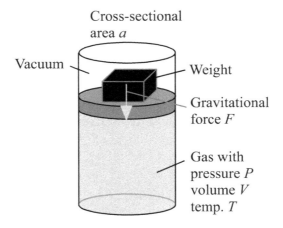

Figure 4.9 A gas enclosed in a cylinder with a freely movable piston. The piston is assumed to run without friction. In the upper part of the cylinder there is a vacuum and a weight is placed on top of the piston. The gravitational force F from the piston and the weight balances the force from the gas pressure on the inside of the piston.

$dV = -a \times ds$, which means that (compare with Equation (4.2))

$$dw = Fds = -\frac{FdV}{a} = -PdV. \tag{4.9}$$

Accordingly, the energy of the gas is changed by[15]

$$dU = dw = -PdV. \tag{4.10}$$

The potential energies of the piston and the weight change in the gravitational field when they move the distance ds, but nothing else is changed for them. Their energy change is $-dw$ because dw is transferred to the gas.

What happens to the total entropy S_{tot} during this change in volume? If the entropy would increase, the volume change would be spontaneous, which implies that the system was not in equilibrium before the volume was changed. Since we assumed that the system was in equilibrium when the volume was V, this cannot be the case. The same applies if the volume would change in the opposite direction (with opposite sign for dV). If the entropy then increases, the

---

[15]For simplicity we disregard the insignificant change in potential energy due to gravity for the gas molecules.

system could not have been at equilibrium at volume V. Thus S_{tot} has a maximal value when the volume is V and for a very small volume change we have $dS_{tot} = 0$, which we can recognize as the general equilibrium condition (see the S_{tot} plot in the upper part of Figure 3.18 – this upper plot applies for the general case). Thus, the entropy is unchanged when the volume is changed by dV (in principle, this applies when dV is very small, in the limit $dV \to 0$).

Now, dS_{tot} consists of the entropy changes of the gas and the surroundings, but the latter change must be equal to zero because the only thing that happened in the surroundings was that the potential energies of the weight and the piston were altered in the gravitational field. Thus, the entropy dS for the gas must be equal to dS_{tot}, but since the latter is zero we have $dS = 0$. Accordingly, if we do a very small change in volume at equilibrium the entropy is unchanged, which is exactly what the second law of thermodynamics says for a reversible process with no heat exchange.

To do a large volume change reversibly, we must perform it in very small steps (infinitely slowly). The gas pressure P will change when the volume is varied and to maintain equilibrium all the time, we must change the force F (by increasing or reducing the weight on top of the piston in small steps) so that $P = F/a$ is fulfilled throughout. The reasoning above applies to each small step and the entropy change is zero,

$$\Delta S = 0,$$

since each step gives zero ($dS = 0$) and ΔS is the sum of changes in all steps.[16]

Note that the gas does not "know" what there is on the other side of the piston; the only essential thing is that there is a force F there that exactly counteracts the force due to the gas pressure all the time. This force F can arise from gravity as in the earlier example, or, for example, from a gas that is located on the exterior side of the piston and that has the same pressure as the gas inside (compare with Figure 4.6).

---

[16]This reasoning applies as an approximation that becomes better when the steps become smaller and smaller, and at the same time, more numerous. When the steps become infinitely small and infinitely many, the reasoning becomes exact. During the volume change, work has been performed, i.e., both V and U have changed. Since $dS = 0$ for each step, we have simply followed a so-called level curve (contour curve) where $S = S(U, V) = $ constant when U and V vary (compare this with a map where each level curve for a mountain indicates the same height above sea level).

The entropy change of the gas in the cylinder is independent of what is causing the force F and it follows that $\Delta S = 0$ provided that the volume change is performed reversibly and $q = 0$.

This reasoning can be generalized to other forms of work, and, generally, the entropy of a system is not changed during **reversible work** provided no heat exchange occurs. If heat dq is added to the system while the reversible work is performed, we have instead $dS = dq/T$ whereby it is only the heat transfer that gives rise to an entropy change.[17] For **irreversible work**[18] we have $dS > dq/T$ and thus $dS_{irrev} > 0$ (see Equations (2.18) and (2.19)).

Note the important difference between heat and work during a reversible process. At thermal equilibrium the temperature is the same for the system and the surroundings, so when heat is transferred reversibly the total entropy is constant since the entropy increase in the system and the entropy decrease in the surroundings (or vice versa) exactly cancel each other. The reason that the total entropy is constant during reversible work is, on the other hand, that the entropy is unchanged in the system as well as in the surroundings, i.e., *individually* in each part.

In classical thermodynamics the second law of thermodynamics is not necessarily formulated as a statement about entropy; there are several equivalent ways to express it. One formulation, usually attributed to Lord Kelvin,[19] is a postulate about conversion of heat

---

[17]When reversible work is performed on the system, one does not excite the system from a lower to a higher energy level, but instead the energy increase occurs by a shift of the energy levels themselves to higher energy values. This does not lead to any change in S. When heat is added, the system is excited to a higher energy level and thereby S increases since the number of ways to distribute the energy is increased.

[18]If work is carried out irreversibly on a system, such as a compression at a rate different from zero, the system becomes excited to a higher energy level simultaneously as the levels are shifted to higher energy values. In the case of an irreversible compression, this results in the addition of a larger amount of energy to the system than during the corresponding reversible compression (i.e., for the same volume change). The excitation that occurs because the rate is not zero leads to an increased S for the same reasons as an addition of heat (i.e., an increase in the number of ways of distributing energy). However, this increase in entropy is not included in the amount dq/T (which is solely the entropy change due to heat transfer), but it is instead included in dS_{irrev}.

[19]It is sometimes also called the Kelvin-Planck formulation of the second law. Lord Kelvin (William Thomson) was a British physicist who lived 1824–1907. He has given important contributions to thermodynamics and other parts of physics. Max Planck (1858–1947) was a German physicist who made many important contributions to theoretical physics, in particular to the basis of quantum mechanics.

to work in cyclically operating machines. That a machine operates cyclically means that it repeats the same process again and again, such as a steam engine or a conventional motor. After one cycle, the machine will return to the same macroscopic state as it had at the beginning of the cycle. Kelvin's formulation of the second law of thermodynamics states:

> It is not possible to construct a cyclically operating machine that takes heat from a heat reservoir at one temperature and converts it completely into work without changing something else permanently.

(A heat reservoir is a system at a certain constant temperature from where one can take heat.)

It is an intellectual challenge to show that Kelvin's formulation is equivalent to the following formulation of the second law, which corresponds to the one we have used:

> There is a state function called entropy, S, which has the property that $dS = dq/T + dS_{irrev}$, where dq is added heat, T is the absolute temperature, and $dS_{irrev} \geq 0$, where the equal sign applies for reversible processes.

In this book, we will not show that these formulations are equivalent; the arguments which show that Kelvin's formulation leads to the existence of entropy and gives its properties can be found in many textbooks in thermodynamics (it is usually in this rather abstract and difficult way that entropy is introduced into the traditional teaching of thermodynamics).

However, it is not so difficult to show that Kelvin's formulation is a correct statement based on what we know about entropy (i.e., from the formulation of the second law that we have used). Because the process performed by the machines is cyclic, this means that all properties of the system (the machine) are equal before and after a cycle, for instance $U_{before} = U_{after}$ and $S_{before} = S_{after}$, which means that $\Delta U = 0$ and $\Delta S = 0$. Since $\Delta U = q + w$ (according to the first law of thermodynamics), we have $q + w = 0$, that is, $q = -w$. This means that all the energy in the form of heat q that is taken up by the machine from the heat reservoir must be delivered in the form of work to the surroundings (delivered work $= -w$), as illustrated in Figure 4.10. The question is whether this can happen or not.

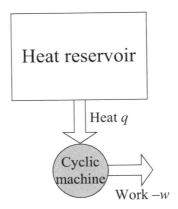

Figure 4.10　A hypothetical machine that operates cyclically and that takes up the heat q from a heat reservoir and delivers work $-w$ to the surroundings during a cycle.

If the machine should be able to do work (that is, $w < 0$), then q must be positive. The entropy change of the system due to the uptake of heat q is necessarily positive (the heat $q > 0$ absorbed at temperature $T > 0$ gives the entropy contribution q/T). The only additional entropy change that the machine can have is from the contribution ΔS_{irrev}, which *cannot* be negative. If $q > 0$, this means that ΔS for the machine during a full cycle consists of a positive contribution and a contribution that is not negative. Thus, it is impossible to have $\Delta S = 0$, which is required for the process to be cyclic. The only possibility to obtain $\Delta S = 0$ is that $q = 0$ and that the process is reversible ($\Delta S_{irrev} = 0$). This implies that $w = 0$, and hence the machine cannot perform any work. This is what Kelvin's formulation of the second law of thermodynamics expresses.

A **perpetual motion machine** is a machine that produces work without any energy being supplied. Such a machine is impossible to construct according to the first law of thermodynamics, because energy cannot be created out of nothing but must be supplied from somewhere else. A perpetual motion machine of the second kind is a machine that produces work by completely converting heat into work without anything else being changed. Such a machine is impossible according to the second law of thermodynamics. If it were possible to make one, you could, for example, have a ship that is driven forward

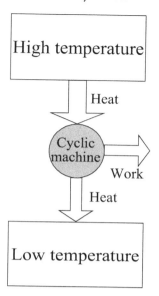

Figure 4.11   A heat engine takes energy in the form of heat at a high temperature, converts some of it to work (performed on the surroundings), and delivers the rest of the energy as heat at a low temperature.

without fuel by taking heat from the ocean and converting it to work that drives the propeller.

However, it is possible to construct a machine that produces work by *partially* converting heat into work, such as a steam engine. It takes energy as heat at a high temperature (the hot steam) and delivers a smaller amount of heat at a low temperature (the cold surroundings). The energy difference is the work that the steam engine performs. Such a kind of machine that produces work by taking heat at high temperature and releasing part of it at low temperature is generally called a **heat engine**, schematically depicted in Figure 4.11.

A heat engine which is run in the opposite manner, so that it uses work to take heat from a low temperature and deliver a greater amount of heat at a high temperature, is a heat pump that we discussed earlier in Section 4.1. This principle is also used in refrigerators (Figure 4.1).

Key points

- The second law of thermodynamics, $dS = dq/T + dS_{irrev}$, where $dS_{irrev} \geq 0$, implies that the entropy of a system can be changed only by heat transfer ($q \neq 0$) or by the occurrence of an irreversible process ($dS_{irrev} > 0$).

- Reversible work does not give rise to any entropy change.

- The second law of thermodynamics can alternatively be formulated: It is not possible to construct a cyclically operating machine that takes heat from a heat reservoir at one temperature and converts it completely into work without changing something else permanently.

4.4   The gas follows the law
The ideal gas law

Let us examine an ideal gas that is expanded without any heat exchange with the surroundings. We assume that the expansion takes place reversibly. As we have seen, when the volume of the gas increases the number of possible particle configurations grows and this results in a positive contribution to the entropy, ΔS_{conf}, given by Equation (4.8). Upon expansion, the gas performs work on the piston which causes the energy of the gas to decrease. This energy loss implies a negative contribution to the entropy of the gas, denoted ΔS_{ener}, because there is less energy to distribute among the gas molecules after the expansion. Since reversible work does not give rise to any change in the entropy of the gas, these two contributions must exactly cancel out. This observation has an important consequence as we shall now show, namely, the *ideal gas law*. Anyone who is not interested in details can go directly to Equation (4.14), but it is a good idea to note the principles in what follows. We first calculate the entropy contributions dS_{conf} and dS_{ener} when the volume is changed by a small amount dV. (These contributions are given by Equations (4.11) and (4.12)). From the condition $dS = dS_{conf} + dS_{ener} = 0$, expressed in Equation (4.13), then follows the result, Equation (4.14).

A LITTLE DERIVATION*
Let us examine the different contributions to the entropy change for the expansion. When we do an expansion with a small volume

change dV, we have $V_{\text{before}} = V$ and $V_{\text{after}} = V + dV$. The entropy contribution from the increase in the number of configurations, dS_{conf}, becomes, according to Equation (4.8)

$$dS_{\text{conf}} = Nk_B\ln\left(\frac{V + dV}{V}\right) = Nk_B\ln\left(1 + \frac{dV}{V}\right).$$

Since $\ln(1 + x) \approx x$ when x is a small number (as illustrated in Figure 4.12), this means that

$$dS_{\text{conf}} \approx Nk_B\frac{dV}{V}.$$

When dV is a very small number, this approximation applies to a very high accuracy, and we have in practice an equal sign in the

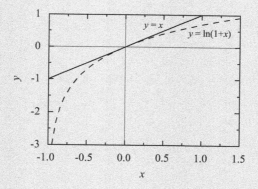

Figure 4.12   A plot of the functions $y = \ln(1 + x)$ (dashed curve) and $y = x$ (solid line), which shows that when x is close to zero, we have $\ln(1 + x) \approx x$. We see that when x is very small, there is practically no difference between $\ln(1 + x)$ and x.

relationship, that is,[20]

$$dS_{\text{conf}} = \frac{Nk_B}{V}dV. \tag{4.11}$$

---

[20]We can alternatively obtain this result mathematically by differentiating S_{conf}.

How large is the change in entropy dS_{ener} because of the energy loss due to the work? The energy change of the gas when the gas molecules push the piston during the expansion is given by Equation (4.10)

$$dU = -PdV.$$

In Section 2.7 we discussed how much the entropy is altered when we vary the amount of energy that is distributed over a system's microstates. In this discussion, we considered the variation in energy due to heat transfer, but the result is more general than that. We saw that the derivative dS/dU is the rate of entropy change when energy is varied, and according to our definition of temperature we have $dS/dU = 1/T$ (Equation (2.14)). If we multiply this derivative with the energy amount dU, we obtain the change in entropy (Figure 4.13)

$$dS_{\text{ener}} = \frac{1}{T}dU$$

and when we insert dU from above we obtain

$$dS_{\text{ener}} = -\frac{P}{T}dV. \tag{4.12}$$

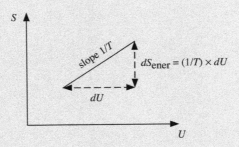

Figure 4.13   The entropy S plotted as a function of the energy U (solid line) has the derivative $1/T$ (= the slope of the curve). When U increases with dU (= the base of the right-angled triangle) S accordingly increases with the slope × the base.

The entropy change dS for the gas is given by the sum of these two contributions[21]

$$dS = dS_{conf} + dS_{ener}.$$

When we insert Equations (4.11) and (4.12), and the condition $dS = 0$, we obtain

$$dS = \left(\frac{Nk_B}{V} - \frac{P}{T}\right)dV = 0. \tag{4.13}$$

Since $dV \neq 0$, the parentheses must be equal to zero, that is, $Nk_B/V = P/T$, which can be written as $PV = Nk_BT$. This is our final result.

The **ideal gas law** reads

$$PV = Nk_BT. \tag{4.14}$$

It is a relationship between the number of molecules, pressure, volume, and absolute temperature, which applies generally to an ideal gas. Alternatively, we can express the ideal gas law in terms of the number of moles $n = N/N_{Av}$ and the universal gas constant $R = k_B N_{Av}$

$$PV = nRT. \tag{4.15}$$

We are now going to examine some of its consequences.

If we divide Equation (4.14) by P on both sides, we obtain $V = bT$ where $b = Nk_B/P$. Thus, the gas volume is proportional to temperature when the pressure and the number of gas molecules are constant (whereby b is constant). This applies, for instance, to the gas in the container in Figure 4.9 where P is constant since $P = F/a = $ constant. The distance between the piston and the container bottom is proportional to the gas volume and thus also proportional to the temperature. By measuring this distance, we can determine the temperature

---

[21] For the mathematically versed it can be said that dS is given by the differential expression

$$dS = \left(\frac{\partial S}{\partial V}\right)_{U,N} dV + \left(\frac{\partial S}{\partial U}\right)_{V,N} dU,$$

where the first term on the right-hand side yields dS_{conf} and the second dS_{ener} where $(\partial S/\partial U)_{V,N} = 1/T$ (the subscript on the derivative shows which variables are held constant in the partial derivation). Note that we implicitly assume that the number of particles N is constant throughout the current section.

of the gas when we know how many moles of gas the container contains.[22]

This relationship can be used to construct a thermometer. By placing the container with the piston (= the thermometer) in contact with some system so that heat can be exchanged, we can determine the system's absolute temperature by measuring the gas volume in the thermometer. We assume thereby that the thermometer is so small that the heat exchange does not significantly affect the temperature of the system. At thermal equilibrium the temperature of the thermometer and the system is the same. If we know the force F, how much gas we have in the cylinder and the value of k_B, we can mark the cylinder with a scale in Kelvin units[23] and then directly read the absolute temperature from the piston position.

A similar principle is used, for example, for ordinary alcohol thermometers, but then it is the volume of the liquid (the alcohol) that is read on a scale which has been calibrated in advance. In contrast to the case of an ideal gas, the liquid's volume cannot be related in a simple manner to the absolute temperature. Instead, one has to do the calibration empirically by comparing with other thermometers or by using freezing/boiling points of a substance (for example, water) as points of reference for some scale. This is the basis for so-called empirical temperature scales such as the Celsius and Fahrenheit scales.

By dividing Equation (4.14) with V on both sides, we obtain

$$P = \rho k_B T, \tag{4.16}$$

---

[22]If the distance between the piston and the container bottom is h, we thus have $T = \text{const} \times h$ with $\text{const} = Mg/Nk_B = Mg/nR$, where M is the combined mass of the weight and the piston, and g is the acceleration due to gravity. We assume of course that the gas to a good approximation behaves like an ideal gas. This means that the temperature cannot be too low. (If one is to determine the temperature with high precision it is necessary to take into account that an actual gas is not ideal – a complication we do not bother with here.)

[23]In this context, we can clearly see the connection between the unit of temperature and the value of k_B. Nothing fundamentally prevents us from choosing any value for k_B; we would even be able to choose its value equal to 1. If we did the latter, we would have entropy as a dimensionless number (i.e., $S = \ln \Omega$) and measure the absolute temperature in energy units (from $T = [dS/dU]^{-1} = dU/dS$ when V and N are constant, i.e., $T = (\partial U/\partial S)_{V,N}$). It is only when we insist on measuring temperatures in a unit equal to $1/100$ of the difference between the boiling point and freezing point of water at about atmospheric pressure at the Earth's surface that k_B acquires the value we usually use (and there is obviously nothing fundamental with that choice).

where $\rho = N/V$ is the molecular density of the gas, that is, the number of molecules per unit volume. Equation (4.16) implies that the gas pressure is proportional to the density when the temperature is constant. A doubling of the density thus gives a doubling of pressure. This is because twice as many molecules collide per unit time against the walls of the container for the gas; the total force against the walls is therefore doubled.

Equation (4.16) also implies that the gas pressure is proportional to the temperature when the density is constant. A doubling of the temperature of the gas thus gives a doubling of the pressure. This increase in P is a consequence of the fact that the molecules move faster when the temperature is increased, and therefore they collide more strongly and more frequently with the container walls. The question is, how much faster do they move (on average)?

One might think that the pressure is proportional to the average molecular speed, but this is not the case. Instead, it is the average kinetic energy (translational energy) per molecule, $\bar{\varepsilon}_{tr}$, that matters (a bar over a symbol indicates average value). By examining the molecular collisions with a wall, one can show that (see Appendix D) the pressure is given by

$$P = \frac{2\rho}{3}\bar{\varepsilon}_{tr}. \tag{4.17}$$

Since the translational energy of a molecule with speed v and mass m is given by $\varepsilon_{tr} = mv^2/2$, a doubling of $\bar{\varepsilon}_{tr}$ corresponds to a doubling of the value of v^2 averaged over all molecules. The speeds of the molecules will on average increase by a factor of $\sqrt{2}$ when the temperature and thereby the pressure is doubled.[24]

If we compare Equations (4.16) and (4.17) we see that, $k_B T = 2\bar{\varepsilon}_{tr}/3$, that is,

$$\bar{\varepsilon}_{tr} = \frac{3}{2}k_B T, \tag{4.18}$$

which means that the average translational energy of the molecules and the temperature are proportional to each other. This provides a direct illustration of the fact that if one somehow increases the kinetic energy, the temperature will rise, and vice versa. It also provides an illustration of the fact that an energy amount of the order $k_B T$ is easily

---

[24]One can alternatively show that this applies by taking the average of $|\mathbf{v}|$ for all molecules.

accessible to a molecule at temperature T (see the discussion on the availability of energy in Section 2.8).

A, perhaps, astonishing fact is that Equation (4.18) is not only valid for ideal gases, but it also applies for strongly interacting molecules in a liquid[25] (a more in-depth analysis is needed to show this). The relationship between temperature and motion is accordingly quite general. The other results in this section are, however, valid for ideal gases only.

For a monatomic ideal gas, the translational energy constitutes the entire energy, so for N molecules we have $U = N\bar{\varepsilon}_{tr}$.[26] By inserting $\bar{\varepsilon}_{tr}$ from Equation (4.18) we obtain

$$U = \frac{3}{2}Nk_BT = \frac{3}{2}nRT \quad \text{(monatomic ideal gas),} \tag{4.19}$$

where n is the number of moles of gas. For an ideal gas of polyatomic molecules, the energy is larger than this at a given temperature, because such molecules have energy also in the form of, for instance, vibrational and rotational energies. These energy contributions increase too as the temperature increases, and the molecules vibrate and rotate faster.

Equation (4.18) is an example of the general fact that the average energy per particle, $\bar{\varepsilon}$, for an ideal gas only depends on temperature, $\bar{\varepsilon} = \bar{\varepsilon}(T)$. This is because the molecules in an ideal gas do not interact with each other, so the energy is independent of the distance between the molecules, that is, independent of the gas density. The total internal energy of the ideal gas is $U = N\bar{\varepsilon}(T)$.

Key points

- The ideal gas law, $PV = Nk_BT$, relates the pressure P, volume V, and temperature T for an ideal gas of N particles. The law can also be written $PV = nRT$, where n is the number of moles of gas.

---

[25]The equation applies if the motions of the molecules are described by Newtonian mechanics, which is a good approximation for many liquids. In cases where quantum mechanics describe the motions of the molecules, the relationship between temperature and movement is more complicated.

[26]We assume here that the molecules are in their electronic ground state, which applies if the temperature is not too high. The energy of the ground state of the gas is here taken as the zero for the energy scale. If some other zero is used for this scale, $U = N(\bar{\varepsilon}_{tr} + \varepsilon_0)$ where ε_0 is the energy of the ground state of a molecule.

- A molecule has an average translational (kinetic) energy, $\bar{\varepsilon}_{tr}$, that is proportional to the temperature.

- The energy U of an ideal gas with N molecules is $U = N\bar{\varepsilon}(T)$, where the average energy per particle, $\bar{\varepsilon}$, depends only on temperature. For a monatomic ideal gas $\bar{\varepsilon} = \bar{\varepsilon}_{tr}$, while for ideal gases of polyatomic molecules $\bar{\varepsilon} > \bar{\varepsilon}_{tr}$ since such molecules have, for example, vibrational and rotational energies in addition to the translational one.

4.5   To heat the kettle
Heat capacity

When we heat a kettle of water on the stove, we may be interested in how much energy is needed to raise the temperature, for example, from 20 to 100°C. (At least we should be if we want to save energy.) In order to do this as cheaply as possible (and as quickly as possible) we should make sure that a lid is tightly closed so very little energy is used to evaporate water. It will be most efficient if we use a closed vessel, so practically all the energy is utilized to heat the liquid and nothing else.

The amount of energy used is proportional to the temperature increase ΔT (at least when ΔT is sufficiently small, which we assume here) and the quantity of heat q we need to add to the system is given by

$$q = C\Delta T, \tag{4.20}$$

where C is a proportionality constant. C is called **heat capacity** and it depends on the amount of substance to be heated: to warm up 2 liters of water a certain number of degrees requires twice as much energy than to warm up 1 liter the same number of degrees, so C for 2 liters is twice that for 1 liter. The value of C depends also on what substance we heat. More heat is needed to raise the temperature of 1 mole of water by ΔT than to do it for one mole of iron, so C is greater for water than for iron.

If we take $\Delta T = 1 K (= 1°C)$, we see from Equation (4.20) that $q = C$. Thus, the heat capacity is the amount of heat needed to raise the temperature of the system by one degree. The term heat capacity refers to the system's ability to "store" energy added as heat. The larger the C,

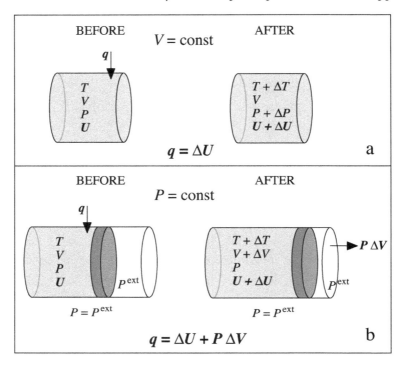

Figure 4.14   Addition of heat q at (a) constant volume, V = const, and (b) at constant pressure, $P = P^{\text{ext}}$ = const. In case (a) all added energy (q) remains in the system and is used to increase the energy of the molecules there, ΔU. In case (b), a part of the added energy is used to push the piston outwards against the external pressure, whereby the energy of the molecules in the surroundings is increased. Only a part of the added energy will remain as an increased energy of the molecules in the system, ΔU.

the more energy can be stored for a given temperature increase. The same amount of heat can subsequently be removed from the system during the corresponding temperature decrease (we then have negative q and ΔT).

Let us, for example, warm up a system that is enclosed in a vessel, the walls of which are such that the volume V of the system is constant (a pressure cooker is approximately such a vessel). When we add energy in form of heat q, the energy of the system will increase by $\Delta U = q$ (Figure 4.14a). No work is performed during the heating,

$w = 0$, because the volume is constant. Thus, we have

$$q = C_V \Delta T = \Delta U \qquad \text{when } V = \text{constant}, \qquad (4.21)$$

where C_V is the heat capacity according to the definition in Equation (4.20). We have set the subscript V on C_V to emphasize that the relationship $q = C_V \Delta T$ applies at constant volume.

If we have a monatomic ideal gas in the vessel, Equation (4.19) applies and we accordingly have

$$\Delta U = \frac{3}{2} N k_B \Delta T \quad \text{(monatomic ideal gas)}, \qquad (4.22)$$

so the proportionality constant is $C_V = \frac{3}{2} N k_B = \frac{3}{2} nR$ in this case. All added heat is used to increase the speeds of the molecules (increased translational energy $N \bar{\varepsilon}_{tr}$) and thereby the temperature. Since the density $\rho = N/V$ is constant, the gas pressure P is raised when T increases (this follows from Equation (4.16)). As we have seen, the rise in pressure occurs because of the increased speeds of the molecules.

If the gas instead consists of an equal number of polyatomic molecules, only a fraction of the added energy q will go to an increase in the speeds of the molecules (increased translational energy). The rest will mainly go to an increase in their vibrational and rotational motions.[27] If we want to raise the temperature equally as much as for the monatomic case, we therefore need to add more heat (note that $\bar{\varepsilon}_{tr}$ must be increased equally in both cases for an equal temperature rise). Thus, the heat capacity C_V for a gas of polyatomic molecules is larger than for a gas of monatomic molecules.

As a further example, we take a system that has constant pressure. It can, for instance, be contained in a vessel like the cylinder in Figure 4.6, where the system pressure P at equilibrium is equal to the external pressure, $P = P^{ext} = \text{constant}$ (for example, a constant atmospheric pressure). For the pressure to remain constant when we heat up the system, the volume V must increase (Figure 4.14b). This increase will make the density ρ of the system decrease and the pressure P will remain unchanged despite the rise in temperature. The process can be described as follows: When heat is added to the system, its molecules take up energy and, among other things, they increase their speeds. In doing so, they expose the piston to a slightly larger force from the

---

[27]When the temperature becomes high, the electronic states of the molecules are also involved for both monatomic and polyatomic molecules.

inside than the force on the outside due to P^{ext}. The molecules there-fore push the piston outwards (against the external pressure) and V increases. As they thereby make the piston move, they lose some of the energy that they have obtained. This means that their net increase in energy is smaller than if the piston had been fixed (as it was in the case of constant volume). The volume increase continues until the density has decreased so much that P again is equal to the external pressure P^{ext} (despite the fact that the molecules in the system now move faster). The rise in P over the value P^{ext} is insignificant if we make the process slow enough, and in practice we can consider P to be constant, $P = P^{\text{ext}}$, all the time. During the expansion, the system performs work equal to $P\Delta V$ on the surroundings (the work *on the system*, $w = -P\Delta V$, is negative). This is the energy that the molecules of the system give up in order to move the piston against the external pressure (Figure 4.14b).

Since the increase in the molecules' energy is less than when the volume was constant (for a given added heat q), the temperature rise ΔT is smaller. The question is, how much smaller? How large is the heat capacity at constant pressure? As we have seen, the added en-ergy q goes on one hand to raise the energy of the molecules in the system, ΔU, and on the other hand to carry out the work $P\Delta V$ on the surroundings. Accordingly, $q = \Delta U + P\Delta V$ and we have (compare with Equation (4.21))

$$q = C_P\Delta T = \Delta U + P\Delta V \qquad \text{when } P = \text{constant,} \qquad (4.23)$$

where C_P is the heat capacity according to the definition in Equation (4.20). We have set the subscript P on C_P to emphasize that the re-lationship $q = C_P\Delta T$ applies at constant pressure. Compared to the case with constant volume, it follows that $C_P > C_V$ since the temper-ature rise for a given supplied heat q is smaller at constant pressure. (If the temperature increase is to be the same as for constant volume, we must therefore add more heat.)

A LITTLE DERIVATION*

The difference between C_P and C_V can be easily determined for an ideal gas. From the ideal gas law (4.14) follows that the term $P\Delta V$ in

Equation (4.23) can be written

$$
\begin{aligned}
P\Delta V &= PV_{\text{after}} - PV_{\text{before}} \\
&= Nk_B T_{\text{after}} - Nk_B T_{\text{before}} = Nk_B \Delta T.
\end{aligned}
\tag{4.24}
$$

Let us first consider a monatomic ideal gas as an example. Equation (4.22) says that $\Delta U = \frac{3}{2}Nk_B\Delta T$ in this case, which, as we have seen, can be expressed as $\Delta U = C_V\Delta T$ with $C_V = \frac{3}{2}Nk_B$. Hence, $\Delta U + P\Delta V = C_V\Delta T + Nk_B\Delta T$, where we have inserted $P\Delta V$ from Equation (4.24). The last equality of Equation (4.23) can therefore be expressed as

$$
C_P\Delta T = C_V\Delta T + Nk_B\Delta T,
$$

which implies that $C_P = C_V + Nk_B$ since $\Delta T \neq 0$. We accordingly have

$$
C_P = C_V + Nk_B = C_V + nR.
\tag{4.25}
$$

For the monatomic ideal gas we have $C_P = \frac{3}{2}Nk_B + Nk_B = \frac{5}{2}Nk_B$.

Equation (4.25) is, in fact, always valid for ideal gases, not only for monatomic ones. The reason for this is, as mentioned in Section 4.4, that the average energy per molecule in an ideal gas depends only on temperature, $\bar{\varepsilon} = \varepsilon(T)$, and is the same irrespective of the gas density. Therefore, when the temperature is varied between two values, the internal energy of an ideal gas, $U = N\bar{\varepsilon}(T)$ changes equally irrespective of whether the pressure or the volume is constant. Thus, $\Delta U = C_V\Delta T$ is valid for an ideal gas both in the cases of constant volume and constant pressure. When we insert $\Delta U = C_V\Delta T$ and Equation (4.24) in the last equality of Equation (4.23), the result, Equation (4.25), follows in the same manner as above.

To conclude, the heat capacity at constant pressure is greater than at constant volume, because only a part of the added heat goes to the raising of the temperature. The rest goes to an increase in the volume. If one wants to achieve a temperature rise of, say, 10 degrees, one thus has to add more heat to a system at constant pressure than at constant volume. It is this fact that the larger heat capacity describes.

Key points

- The heat capacity C of a system specifies how much heat should be added to raise the temperature by one degree.

- To change the temperature by ΔT, the heat $q = C\Delta T$ is needed (provided that ΔT is not too large).

- If the volume is kept constant, the heat capacity is denoted C_V and if the pressure is kept constant C_P. In the former case $q = C_V\Delta T$ and in the latter $q = C_P\Delta T$.

- When the pressure is held constant, the volume is increased during the addition of heat and a larger amount of heat is required to raise the temperature by ΔT than when the volume is constant. A portion of the added energy goes to the expansion of the system (that is, to the work done on the surroundings) instead of raising the temperature. Therefore $C_P > C_V$.

4.6   The balance of two bank accounts
The concept of enthalpy

In the previous section we found that the temperature of a system is increased more if we add an amount of heat q at constant volume than at constant pressure. At constant volume, all heat goes to an increase in the energy of the molecules in the system (Equation (4.21) and Figure 4.14a)

$$q = \Delta U \qquad \text{when } V = \text{constant,} \tag{4.26}$$

while at constant pressure, some of the added heat goes to an increase of the volume and only a portion goes to an increase in the energy of the molecules (Equation (4.23) and Figure 4.14b)

$$q = \Delta U + P\Delta V \qquad \text{when } P = \text{constant.} \tag{4.27}$$

The amount of energy $P\Delta V$ is thereby transferred to the surroundings in the form of work during the volume increase and only the energy ΔU remains in the system. If q is equal in Equations (4.26) and (4.27), ΔU in the former is larger than in the latter.

Let us now make the system return to its original state by removing the same amount of heat from it. Thereby the whole of this energy is taken from the molecules of the system in the first case, while the energy will be drawn from both these molecules and from the surroundings in the second case. The absolute values of q, ΔU, and ΔV are thereby the same as during the addition of heat, but the signs are reversed (negative). Let us look at this process for the case of constant pressure, such as when the system is enclosed in a container like in Figure 4.6. (The following reasoning is similar to that for the addition of heat discussed in the previous section, so it suffices with a short version here.) When we take the heat from the system, the speeds of the molecules decrease and the pressure P becomes slightly less than the external pressure P^{ext}. The surroundings will therefore push the piston inwards, whereby the molecules in the system receive a speed increment. This increment makes the speed reduction due to the heat removal less than in the case of constant volume, where all energy was taken from the molecules of the system. At constant pressure, we accordingly take out a larger amount of heat than what corresponds to the net decrease in the molecular energies. The energy difference comprises the work that the surroundings do on the system via the piston movement.

When we add heat for the case of constant pressure, the energy is "stored" partly in the molecules of the system and partly in the surroundings. We can take the heat back out, and the energy is then being taken partly from the system and partly from the surroundings. It is as if we deposit money in a bank, but distribute the amount between two different bank accounts. When we take the money back out, we take it from both accounts. That the bank has placed money from one account "abroad" (in the surroundings) does not play any role for our overall balance. The bank takes the money back again from abroad when we need it. What is interesting in this context is the total balance, and it is the same way in thermodynamics. From Equation (4.27) follows

$$
\begin{aligned}
q &= \Delta U + P\Delta V = U_{after} - U_{before} + PV_{after} - PV_{before} \\
&= U_{after} + PV_{after} - (U_{before} + PV_{before}) = \Delta(U + PV)
\end{aligned}
$$

when P = constant.

We see that the quantity of heat is equal to the change in $U + PV$, which hence plays the role of the "total balance" at constant P. This combination of variables has an important role to play and it is therefore handy to introduce a separate symbol H for it,

$$H = U + PV, \tag{4.28}$$

and to give it a name, **enthalpy**. We accordingly have

$$q = \Delta H \qquad \text{when } P = \text{constant}, \tag{4.29}$$

and H is constructed so that it keeps track of the bookkeeping of the two "accounts" (the system and the surroundings) which are relevant for heat transfer at constant pressure.

In the case of constant volume, we have no access to the second "account" (the surroundings), so we "place" all heat as increased energy of the molecules in the system (Equation 4.26). Remember that U, the internal energy, is the total energy of the molecules, that is, the sum of all kinetic energy (translational, vibrational, and rotational energy) and potential energy (interactional energy) of the molecules in the system.

When the pressure is constant we have from Equation (4.28) that

$$\begin{aligned} \Delta H &= \Delta(U + PV) = (U_{\text{after}} + PV_{\text{after}}) - (U_{\text{before}} + PV_{\text{before}}) \\ &= U_{\text{after}} - U_{\text{before}} + PV_{\text{after}} - PV_{\text{before}} = \Delta U + P\Delta V. \end{aligned}$$

Therefore, we can write Equation (4.23) as

$$q = C_P \Delta T = \Delta H \qquad \text{when } P = \text{constant}, \tag{4.30}$$

which we can compare with Equation (4.21). The heat capacity at constant pressure is thus directly related to the variation in enthalpy during temperature changes.

We saw in Section 4.4 that the average energy per molecule for an ideal gas depends only on temperature, $\bar{\varepsilon} = \bar{\varepsilon}(T)$. This is a consequence of the condition that the molecules do not interact with each other. For a gas of N molecules, this means that the internal energy depends only on temperature, $U = U(T)$. The same applies for the enthalpy, $H = H(T)$, because according to the ideal gas law we have $PV = Nk_BT$. From Equation (4.28) we therefore see that for an ideal gas, $H = U + Nk_BT$, where both terms of the right-hand side depend on T only (when N is constant), and the conclusion $H = H(T)$ follows.

Since $U = U(T)$ and $H = H(T)$ for an ideal gas, it follows that when the temperature is changed from T_{before} to T_{after}, the variations

ΔU and ΔH for the gas are determined solely by these two temperatures, *irrespective* of what happens to the pressure and the volume. Therefore,

$$\Delta U = C_V \Delta T \quad \text{for ideal gas}$$
$$\Delta H = C_P \Delta T \quad \text{for ideal gas} \tag{4.31}$$

when the particle numbers are unchanged. These relationships are valid when the pressure is constant, the volume is constant, or both P and V vary. Note, however, that the relationship $q = C_V \Delta T$ is valid *only* for constant volume and $q = C_P \Delta T$ *only* for constant pressure.

Key points

- The heat added to a system when volume is constant (no work is done) is equal to the change in the system's internal energy, $q = \Delta U$.

- When the pressure is held constant, the volume is increased during addition of heat and the added energy q goes partly to an increase in energy of the molecules in the system and partly to the expansion of the system.

- The enthalpy, $H = U + PV$, is a system property that is so designed that when heat q is added, H keeps track of both the energy that stays in the system and the energy that goes out into the surroundings in the form of work during the expansion. H changes by both amounts and thus $q = \Delta H$ when the pressure is constant.

- The enthalpy H and the internal energy U for an ideal gas depend only on the temperature (provided no particle numbers are changed), $H = H(T)$ and $U = U(T)$.

4.7 Spontaneity for the most common circumstances
The concept of Gibbs energy

The general criterion for a process to occur spontaneously in a system is, as we have seen, that the total entropy S_{tot} for the system and the surroundings increases. However, when one wants to find out whether a process can occur spontaneously or not, it is convenient to focus solely on the system and not to explicitly consider changes in the surroundings. Therefore, we introduced in Section 3.8 the quantity Helmholtz energy A, which we defined on the basis the system's energy U, entropy S, and temperature T as $A = U - TS$ (Equation (3.6)). A was constructed so that it keeps track of the entropy changes both in the system and the surroundings. We saw that if the process is carried out at constant temperature (because the system can exchange heat freely with a thermostat at temperature T) and constant volume, A decreases for a spontaneous process. The relationship between the change in A and S_{tot} is, according to Equation (3.7), $\Delta A = -TS_{tot}$ when T and V are constant, so a decrease in A represents an increase of S_{tot}, as illustrated in Figure 3.18. The equilibrium condition at constant T and V is, as we have also seen, that A has reached a minimal value and thus $dA = 0$.

Processes at constant temperature and volume are, however, not very common in practice.[28] Instead, processes *at constant temperature and pressure* are more common and therefore more important. Under these conditions, it is not A that can be used to determine whether or not a process can occur spontaneously and whether equilibrium is obtained. However, the general criterion $\Delta S_{tot} > 0$ for spontaneous processes and the equilibrium condition $dS_{tot} = 0$ still apply. Also in this case, it would, however, be helpful to have a criterion that only uses some property of the system itself, like the quantity A above.

To find such a quantity, we will argue in a similar manner as we did when introducing A in Section 3.8. At the beginning of the section, we derived Equations (3.1) and (3.2), which apply generally provided the temperature is constant. Equation (3.2) says that a spontaneous process is characterized by $\Delta S - q/T > 0$, where q is the heat delivered to the system to keep T constant during the process. When the pressure is constant, we have seen in the previous section (Equation (4.29)) that $q = \Delta H$, so we obtain (compare with Equation (3.3))

$$\Delta S - \frac{\Delta H}{T} > 0 \qquad \text{(for spontaneous process at constant } T \text{ and } P\text{).}$$
$$(4.32)$$

As an illustration of this equation, let us discuss an endothermic process, i.e., a process where $q = \Delta H > 0$. Since $\Delta S_{surr} = q_{surr}/T =$

---

[28]Important examples of constant volume cases are batch reactor processes in chemical engineering.

$-q/T = -\Delta H/T$, the second term in Equation (4.32) is equal to the entropy change in the surroundings when heat is transferred to the system in order to keep T constant. Equation (4.32) therefore says that an endothermic process is spontaneous provided that the entropy of the system increases more than the entropy of the surroundings decreases, just as before. Thereby, the total entropy S_{tot} increases. Similar reasoning, departing from Equation (4.32), can be carried out for exothermic processes; compare with the discussion of Equation (3.2) in Section 3.8. The fact that Equation (4.32) contains ΔH instead of ΔU, means that it has taken into account that the volume of the system may change during the process when P is constant.

According to Equation (3.1), $\Delta S_{tot} = \Delta S - q/T$ and hence we have

$$\Delta S_{tot} = \Delta S - \frac{\Delta H}{T} = -\frac{\Delta H - T\Delta S}{T} \quad \text{(at constant } T \text{ and } P\text{).} \quad (4.33)$$

This equation says that the total entropy change in the current case can be expressed in the enthalpy and entropy changes of the system. Like in the passage from Equation (3.4) to (3.5) in Section 3.8, we obtain from Equation (4.33)

$$\Delta S_{tot} = -\frac{(H - TS)_{after} - (H - TS)_{before}}{T} \quad \text{(at constant } T \text{ and } P\text{).}$$

$$(4.34)$$

The total entropy change is thus proportional to the change in the value of $H - TS$. (In the case of constant volume it was instead $U - TS$, namely, A, that had this role.) It is therefore convenient to introduce a symbol for $H - TS$ and give it a name: **Gibbs free energy**

$$G = H - TS \quad (4.35)$$

or simply **Gibbs energy**.[29]

At constant temperature and pressure, it is thus the balance between the system's entropy change ΔS and the effect of its enthalpy change ΔH (in the form of the entropy change in the surroundings) that determines whether a process in the system is spontaneous or not. This balance is expressed in

$$\Delta G = \Delta H - T\Delta S = -T\Delta S_{tot} \quad \text{(at constant } T \text{ and } P\text{);} \quad (4.36)$$

---

[29]The modern, recommended name is Gibbs energy, but Gibbs free energy has long been the commonly used name and is still the most common one in the scientific literature. In this book, the names Gibbs energy and Helmholtz energy are used. Both quantities are, however, called free energies as a collective name.

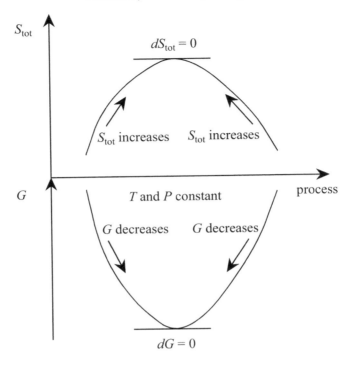

Figure 4.15   A process at constant temperature and pressure can take place spontaneously in the direction of decreasing free energy G (Gibbs energy). Thereby, the total entropy of the system and the surroundings increases. Equilibrium is reached when G has attained its minimum value and at the same time S_{tot} is as large as possible under the given circumstances. At the minimum and maximum point, respectively, the derivative is zero (horizontal line) and $dG = 0$ and $dS_{tot} = 0$.

compare with Equation (3.8). If $\Delta G < 0$, the process can take place spontaneously ($\Delta S_{tot} > 0$), and if $\Delta G > 0$, the reverse process can instead occur spontaneously. The free energy G is thus designed so that it automatically takes into account entropy changes in both the system and the surroundings. This is generally the case for a free energy, and it is this characteristic that gives it such an important role. Depending on the circumstances, there are different free energies (Gibbs energy at constant pressure and Helmholtz energy at constant volume) that have this role.

The properties of G are summarized in Figure 4.15, and we see that equilibrium corresponds to the minimal value of G (compare to

Figure 3.18 and the corresponding discussion in Section 3.8). At the minimum point we have $dG = 0$, which accordingly is the equilibrium condition at constant T and P. Note that the process is spontaneous in one direction to the left of the equilibrium position and in the opposite direction to the right of it.

Since according to Equation (4.28) $H = U + PV$ and according to Equation (3.6) $A = U - TS$, we can write Gibbs energy in Equation (4.35) as

$$G = U + PV - TS = A + PV. \tag{4.37}$$

When the pressure is constant we thus have

$$
\begin{aligned}
\Delta G &= \Delta(A + PV) = (A_{after} + PV_{after}) - (A_{before} + PV_{before}) \\
&= A_{after} - A_{before} + PV_{after} - PV_{before} = \Delta A + P\Delta V.
\end{aligned}
$$

The difference between ΔA and ΔG is therefore $P\Delta V$ at constant P. When gases are released or consumed during the process, the change in volume can be substantial and there is a large difference between ΔA and ΔG. For systems that only consist of solid or liquid substances, however, the volume changes during the process are often small, and then ΔA and ΔG are usually approximately equal. In such cases it is not so important to make a distinction between ΔA and ΔG. (An important exception is when the process is carried out at very high pressure, so the factor P makes $P\Delta V$ large even if ΔV is small.) It is therefore mainly for processes that involve a gas phase that it is important to use Gibbs energy rather than Helmholtz energy when the pressure is constant.

Key points

- **Gibbs energy**, $G = H - TS$, is a system property that is so designed that it keeps track of entropy changes in both the system and the surroundings when T and P are constant.

- When T and P are constant, we have $\Delta G = -T\Delta S_{tot}$.

- The criterion for a process to be spontaneous at constant T and P is that G decreases.

- Equilibrium at constant T and P occurs when G has reached its minimal value and then S_{tot} is as large as possible. At the minimum point for G we have $dG = 0$.

- Depending on the circumstances, different free energies should be used to determine whether or not a process is spontaneous. At constant P and T the Gibbs energy, G, should be used and at constant V and T it is the Helmholtz energy, A, that has this role.

CHAPTER 5

Mixtures and reactions

In this chapter we will make use of what we have learned about entropy, enthalpy, and Gibbs energy in a particularly important application, namely, chemical reactions. When we have a mixture of chemical compounds, a relevant question is whether these compounds can react with each other and form other compounds. If our aim is to produce a certain chemical we would like to know the conditions when it can be formed and preferably the reaction should occur in an optimal manner giving a high yield. In other cases we may want to know under what conditions there is chemical equilibrium in the mixture, so the composition of the mixture does not change with time. With these applications in mind, we will in the following section investigate thermodynamical properties of mixtures of gases. We limit ourselves to ideal gases since we have everything ready for this case, but many of the results have a more general applicability, in particular the *law of mass action* that we derive in Section 5.2. This law, which gives the condition for chemical equilibrium, is a cornerstone in chemistry and was formulated for the first time by Guldberg and Waage[1] in the 19th century on the basis of empirical data.

Properties of mixtures and changes in thermodynamical quantities when mixtures are formed are, however, of general importance. So apart from applications in chemical reactions, this subject is worth a study in itself.

5.1 Take from the bottle and mix
Gas mixtures and standard states

The composition of a mixture can, for example, be specified as the proportion of molecules that are of each kind, such as that 1/3 of the molecules are of species A and 2/3 of species B. This is called the **mole fraction**, x_i, where the subscript i indicates molecular species,

---

[1]Cato M. Guldberg (1836–1902) and Peter Waage (1833–1900) were Norwegian scientists who formulated the law of mass action, also known as the Guldberg-Waage law, for the first time in 1864 and 1867.

whereby in this example $x_A = 1/3$ and $x_B = 2/3$. We have $x_A = N_A/(N_A + N_B) = n_A/(n_A + n_B)$ and analogously for x_B, where N_i is the number of molecules and n_i the number of moles of species i and where the last equality follows from $n_i = N_i/N_{Av}$. The sum of the mole fractions of all components is equal to one, $x_A + x_B = 1$. In the general case $x_i = N_i/N_{tot} = n_i/n_{tot}$, where N_{tot} and n_{tot} are the total number of molecules and moles, respectively. The composition of a mixture can also be given in terms of the molecular density ρ_i of species i, defined as $\rho_i = N_i/V$, or the concentration $c_i = n_i/V$ (number of moles per unit volume). The relationship between them is $\rho_i = c_i N_{Av}$.

A common alternative for gas mixtures is to specify the composition as the **partial pressure** P_i for each component i, defined as $P_i = x_i P$ where P is the total gas pressure (= the sum of all partial pressures). The ideal gas law (4.14) applied to component i becomes

$$P_i V = N_i k_B T \tag{5.1}$$

and written as in Equation (4.16) it is $P_i = \rho_i k_B T$. This implies that the partial pressure is proportional to ρ_i when T is constant. The proportionality can be understood as a consequence of the fact that the number of wall collisions per unit time for species i is proportional to the density of the species.

In Section 2.5 we investigated a case where we mixed two different ideal gases with each other. When the gases are allowed to mix, the number of particle configurations for each particle type increases by a factor of $[V_{after}/V_{before}]^N$, where N is the number of particles of the respective species, V_{before} is the volume that respective gas occupies before mixing, and V_{after} the volume after (see Equation (2.3)). In the example of Section 2.5, $V_{after}/V_{before} = 2$ for both gases. The entropy increase for each gas is $\Delta S = k_B \ln[V_{after}/V_{before}]^N = N k_B \ln[V_{after}/V_{before}]$ (Equation (2.8)), and the total increase in entropy is the sum of the contributions from the two species. The reason why we can simply add the ΔS contributions of the two gases when we mix them is that the molecules of ideal gases do not interact with each other. Each single molecule "does not know about" the existence of the other molecules, irrespective of whether they are of the same or different species. When we make an ideal gas mixture, we can treat each gas independently of the other gases. Let us denote the entropy contribution from a gas of species i as $\Delta S(i)$, for instance, if the gas is carbon dioxide we have $\Delta S(CO_2(g))$.

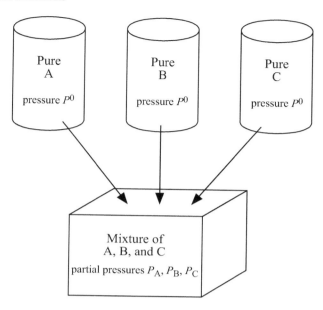

Figure 5.1   The making of a gas mixture from pure gases taken from gas bottles which all have pressure P^0.

Let us now make a mixture of gases by taking a suitable number of molecules from some gas bottles that contain the pure gases and bring the molecules into a container where the gases are mixed. The temperature T is the same the whole time (everything is in contact with a thermostat). Assume that the pure gas in each bottle has pressure P^0, equal for all bottles as illustrated in Figure 5.1. Thus, the gas density is the same in all bottles, $\rho^0 = P^0/k_BT$. Since we can treat each gas separately, we can put them into the container one by one independently of each other. Thereby, the pressure of the gas of species i is changed from P^0 (in the bottle) to P_i (in the container), that is, the final partial pressure of the mixture. How much does the entropy and other quantities of the gases change when we mix them?

A LITTLE DERIVATION

We take N_i molecules (n_i moles) of a pure gas from a bottle. To change the pressure from P^0 to $P_i = N_i k_B T/V$, where V is the volume of the container for the mixture, we need to change the gas density from ρ^0 to the density ρ_i in the mixture, where $\rho_i = N_i/V = P_i/k_BT$.

Note that

$$\frac{\rho^0}{\rho_i} = \frac{P^0/k_B T}{P_i/k_B T} = \frac{P^0}{P_i}.$$ (5.2)

To do this change in density, we must change the volume of the gas from $V_{\text{before}} = N_i/\rho^0$ to $V = V_{\text{after}} = N_i/\rho_i$ and we have

$$\frac{V_{\text{after}}}{V_{\text{before}}} = \frac{N_i/\rho_i}{N_i/\rho^0} = \frac{\rho^0}{\rho_i} = \frac{P^0}{P_i}.$$

Equation (2.8) gives the entropy change when the volume is varied from V_{before} to V_{after}. For the entropy $S(i)$ of the gas of species i we obtain $\Delta S(i) = N_i k_B \ln(V_{\text{after}}/V_{\text{before}}) = N_i k_B \ln(P^0/P_i)$. Let us write $\Delta S(i) = S(i)_{P_i} - S(i)_{P^0}$, where the subscript indicates which pressure the entropy applies for. Thus we have

$$S(i)_{P_i} - S(i)_{P^0} = N_i k_B \ln\left(\frac{P^0}{P_i}\right) = n_i R \ln\left(\frac{P^0}{P_i}\right).$$ (5.3)

Note that the entropy we consider here is the configurational entropy S_{conf}.

How about the enthalpy and the free energy? In Section 4.6, we saw that the enthalpy of an ideal gas depends on the temperature only, $H = H(T)$. Since T is constant, we have

$$H(i)_{P_i} = H(i)_{P^0}$$ (5.4)

so the enthalpy does not change $\Delta H(i) = 0$. From the relationship $\Delta G = \Delta H - T\Delta S$ we can now calculate the change in Gibbs energy and by using Equation (5.3) we obtain $\Delta G(i) = -T\Delta S(i) = -N_i k_B T \ln(P^0/P_i) = N_i k_B T \ln(P_i/P^0)$, where we have inverted the argument of the logarithm and therefore changed the sign in front. Thus

$$G(i)_{P_i} - G(i)_{P^0} = N_i k_B T \ln\left(\frac{P_i}{P^0}\right) = n_i R T \ln\left(\frac{P_i}{P^0}\right).$$ (5.5)

Since the entropy and enthalpy of the mixture can be obtained by adding the contributions from each component, this also applies to Gibbs energy.

We now assume that we know the entropy, enthalpy, and Gibbs energy for each pure gas in the bottle, and we let S^0, H^0, and G^0 denote these quantities where the zero indicates that the pressure of the gas is P^0. More precisely, let 1 mole of the pure gas at pressure P^0 have Gibbs energy $G_m^0(i)$, where subscript m stands for "per mole." For n_i moles, the Gibbs energy is accordingly $n_i G_m^0(i)$, which we denoted $G(i)_{P_0}$ in Equation (5.5).

Gibbs energy for the gas at partial pressure P_i can, according to Equation (5.5), be written

$$G(i)_{P_i} = n_i \left[G_m^0(i) + RT \ln\left(\frac{P_i}{P^0}\right)\right]. \tag{5.6}$$

For a mixture of, for example, n_A moles A and n_B moles B with partial pressures P_A and P_B, respectively, the Gibbs energy is the sum of $G(i)_{P_i}$ for the components, that is, $G(A)_{P_A} + G(B)_{P_B}$, and we obtain

$$G = n_A \left[G_m^0(A) + RT \ln\left(\frac{P_A}{P^0}\right)\right] + n_B \left[G_m^0(B) + RT \ln\left(\frac{P_B}{P^0}\right)\right], \tag{5.7}$$

which we can write as

$$G = n_A G_m^0(A) + n_B G_m^0(B) + n_A RT \ln\left(\frac{P_A}{P^0}\right) + n_B RT \ln\left(\frac{P_B}{P^0}\right). \tag{5.8}$$

The first two terms on the right side of Equation (5.8) is G for the pure gases at pressure P^0. The last two terms is the change in Gibbs energy when the pressures are changed from P^0 to partial pressures P_A and P_B, respectively, in the mixture. Similarly, we introduce $S_m^0(i)$ and $H_m^0(i)$ as S^0 and H^0 per mole of the substance at pressure P^0 (the "molar" entropy and enthalpy, respectively, at P^0) and we obtain the corresponding expressions for entropy and enthalpy. (EXERCISE: Derive the expressions for entropy and enthalpy starting from Equations (5.3) and (5.4).)

The important conclusion from these results is that if we know the entropy, enthalpy, and Gibbs energy per mole of the pure gases at pressure P^0, we have everything we need in order to calculate S, H, and G *for any mixture* of them (assuming that the gases are ideal as a good approximation). This is very convenient because we thereby need only to experimentally determine the data for the pure gases at pressure P^0 (i.e., the gases we have in the bottles) and not for every possible mixture of them.

If we choose $P^0 = 1$ bar $= 10^5$ Pa, the pure gas in the bottle is in the macroscopic state that is commonly referred to as the **standard state** of the substance at the temperature in question (this is actually the meaning of the zero in P^0 and, for example, G_m^0). That the standard state of a substance is the pure substance at 1 bar pressure is a common international convention (agreement). A previous convention has been to instead choose $P^0 = 1$ atm and this choice still occurs in scientific literature and some textbooks. (Note that 1 atm ≈ 1.01 bar so the difference is quite small.) The concept of standard state does not apply only to gases but to all substances regardless of their state of aggregation. Tables of data are available for a large number of substances in their standard state. It is normally specified at which temperature the tabulated value applies (if the temperature is not given, it is usually 25°C). From these data, one can thus calculate S, H, and G for mixtures of the substances (at least when they are ideal gases).

Before proceeding, we should note the following. To specify a value for a thermodynamic property like the internal energy U of a system, one must have agreed on what zero level to use, i.e., the circumstance under which U has the value zero. Experimentally, one always measures differences in energy, for example, the energy of an object when it is on the 10th floor in a building compared to the street level or the energy of a system in one state compared to another state. It may be practical to say, for instance, that the energy of the object is equal to zero when it is at the street level and then one can specify a numerical value of the energy when the object is on the 10th floor (the value is equal to the difference in energy compared to the zero level). When the object is on the 11th floor it has a different (higher) value of energy due to gravity.

However, there is nothing that says that one has to choose the street level as the zero level; perhaps the lowest point of the street is a better choice or the lowest point in the city. If one chooses any of the latter as zero level, one obtains, of course, a different numerical value of the energy. Provided that one is interested in chemical processes, it is reasonable to include the energy for the molecules of the object (i.e., the kinetic and potential energies of the electrons and nuclei in the molecules). Then the energy of the object once again takes on a different value. If one is interested in nuclear processes, it is reasonable to also include the energy of the protons and neutrons in the atomic nuclei. There is simply no obvious zero level and the numerical value of the energy depends on the zero level chosen. However, the energy difference between the 10th floor and the ground floor or between two other states is of course independent of the choice of the zero level.

The same applies to enthalpy and free energy and therefore one has to determine what zero level to use by means of a convention (since U has no natural zero level this also applies to $H = U + PV$ and $G = U + PV - TS$ which contain U). The internationally accepted convention in chemical sciences is:

For a pure element in its most stable form at the temperature in question $H = 0$ and $G = 0$ in the standard state.

Examples of the most stable form of some elements at 25°C are: oxygen (gas), chlorine (gas), bromine (liquid), iodine (solid phase), and carbon (solid phase in the form of graphite).[2]

The enthalpy of a substance is thereby equal to the enthalpy difference between the substance and the elements it consists of. This difference is a quantity called the **enthalpy of formation** and is often denoted $\Delta_f H_m^0$ (subscript f stands for "formation"). For example, the enthalpy of carbon dioxide, $\Delta_f H_m^0(CO_2(g))$, at 25°C is equal to the difference in enthalpy between 1 mole $CO_2(g)$ and the elements it consists of, i.e., 1 mole carbon (in the form of graphite) and 1 mole $O_2(g)$. This enthalpy difference is equal to the enthalpy of the reaction $C(graphite) + O_2(g) \rightarrow CO_2(g)$, that is, the difference in enthalpy between the products and the reactants, $H(products) - H(reactants)$, when one mole of the substance is formed from the elements. This is the reason for the name "enthalpy of formation." Each substance participating in the formation reaction should be in its standard state, as indicated by the superscript 0 on $\Delta_f H_m^0$. Note that for an element in its most stable form, $\Delta_f H_m^0 = 0$, in accordance with the selected zero level. For example, the enthalpy for the reaction $O_2(g) \rightarrow O_2(g)$ is zero since nothing happens and therefore $\Delta_f H_m^0(O_2(g)) = 0$. The same applies for $Cl_2(g)$, but for bromine at 25°C we have $\Delta_f H_m^0(Br_2(g)) > 0$ since the stable form is $Br_2(l)$ and the enthalpy for $Br_2(l) \rightarrow Br_2(g)$ is positive since heat must be added to vaporize $Br_2(l)$.

In this book we will, however, not use the notation $\Delta_f H_m^0$ for the enthalpy of a substance. Instead, we will write H_m^0 and thereby we let

---

[2]Pure carbon in the form of diamond is unstable at 25°C and normal atmospheric pressure. However, the conversion to graphite takes, fortunately, an extremely long time.

the zero level be implicit.[3] The corresponding thing applies to Gibbs energy and we will continue to use the notation G_m^0 for the Gibbs energy of one mole of a pure substance at pressure P^0. Provided that the zero level for G is selected according to the convention above, the *value* of G_m^0 is equal to **Gibbs energy of formation** $\Delta_f G_m^0$, i.e., the difference in Gibbs energy between the products and the reactants when one mole of substance is formed from the elements in their most stable forms. If some other convention would be used for the zero level, G_m^0 would have a different value, but *differences* in Gibbs energy would be the same. Experimentally determined values of $\Delta_f H_m^0$ and $\Delta_f G_m^0$ are tabulated for a large number of substances in their standard state and these values can be used for H_m^0 and G_m^0. (Such tables can be found in various scientific handbooks and in most major textbooks in general chemistry and physical chemistry.)

In contrast, for entropy there exists a natural zero level due to the third law of thermodynamics, namely, $S = 0$ for the substance in perfect crystalline form at absolute zero, $T = 0$. The value of S_m^0 for a substance that is tabulated usually refers to the entropy relative to this zero level. Therefore, even the elements in their most stable forms have values of S_m^0 that are different from zero when $T > 0$. S_m^0 is called the **standard entropy** of the substance.

Key points

- The thermodynamic quantities (for example, S, H, and G) of an arbitrary mixture of ideal gases can be easily calculated if the corresponding properties are known for the pure substances at pressure P^0.

- Thermodynamic properties of many substances in their standard state (the pure substance at pressure $P^0 = 1$ bar) are published in tabular form in various scientific handbooks.

---

[3]This deviation from standard practice is done for pedagogical reasons. For a chemical reaction A $\rightarrow$ B where one mole of A forms one mole of B, the enthalpy of the reaction at constant pressure $P = P^0$ is given by $\Delta H^0 = H_m^0(B) - H_m^0(A)$ in our notation, which conforms to $\Delta H = H_{after} - H_{before} = H(\text{products}) - H(\text{reactants})$ (A and B are in their standard states). The usual way of writing this is $\Delta H^0 = \Delta_f H_m^0(B) - \Delta_f H_m^0(A)$, which is, in the experience of the author, quite confusing for many beginners in the subject. In practice, to find the value of, for example, $H_m^0(A)$ one looks up $\Delta_f H_m^0(A)$ in a table, which is the molar enthalpy for A given the conventional choice of zero level for H.

5.2   Can they react?
Chemical reactions and equilibria

In the previous section we learned how to calculate thermodynamic properties, such as Gibbs energy, for an arbitrary mixture of ideal gases. The only things we need to know in advance are the values of the respective property of the pure gases at pressure $P^0 = 1$ bar (tables are published with such values). If a reaction takes place between the components of the gas mixture, the composition of the mixture will change; the number of reactant molecules will decrease and product molecules will increase. If we keep temperature T and total pressure P constant (the latter by changing the volume of the system if necessary), we know that the process is spontaneous if G decreases. Provided G of the initial state (the mixture of reactants) is higher than G of the final state (the mixture of products), the reaction can occur spontaneously.[4] If, on the other hand, G for the initial state is lower than for the final state, the reaction cannot occur spontaneously but the reverse reaction can. By calculating G for mixtures of gases, we can hence predict whether the reaction is possible or not under the given circumstances ("circumstances" = the current composition of the gas mixture and the given total pressure and temperature).

In many cases the reaction does not go to completion, but we get a mixture of products and reactants as the final result. The reaction will proceed until G has decreased as much as possible (to its minimal value) and then the system has reached equilibrium (compare with Figure 4.15). Since we can calculate G for all possible compositions of the mixture, we can predict the variations in G when the composition changes due to the reaction and thereby determine theoretically at which composition G has minimum. Thus, we can set up the condition for equilibrium – a condition known as the *law of mass action*, which has a central role in chemistry. The current section is primarily concerned with the equilibrium condition and the molecular interpretation of it.

---

[4]The lowering of G is a necessary condition, but in some cases the process may occur so slowly that a progression cannot be detected. One then says that the reaction is *kinetically hindered*. Typically, for such a reaction a very large energy barrier has to be passed between the initial and the final states, which makes the probability to pass the barrier very low even if the final state has a lower free energy than the initial one. The speed of the reaction can be increased by, for example, the addition of a catalyst that lowers the barrier.

5.2.1   Equilibrium of type $A \rightleftharpoons B$

We start with a simple example, a mixture of two isomers A and B in the gas phase which can be converted into one another

$A \rightleftharpoons B,$

for example, A = cis-butene and B = trans-butene as shown in Figure 5.2. They are the two possible forms of 2-butene.

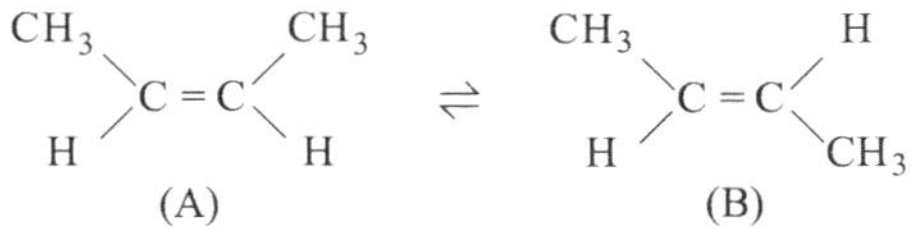

Figure 5.2   Equilibrium between cis-butene (A) and trans-butene (B).

The conversion between the cis and trans forms is achieved by rotation around the double bond. Of the two isomers, cis-butene has the highest energy for spatial reasons; the two methyl groups are quite bulky and repel each other as they sit on the same side of the molecule. The hydrogen atoms do not take up as much space, so a hydrogen atom on the same side as a methyl group does not compete for space. Therefore, trans-butene has the lowest energy. A sketch of how the energy depends on the angle between the methyl groups is shown in Figure 5.3. There is a fairly high energy barrier between the two isomers that must be overcome to enable conversion between these two forms. The essential matter for the equilibrium properties is, however, mainly the difference in energy between the two minimum points.

The lower energy of trans-butene implies, as we shall see, that this isomer is normally favored. For $T = 400$ K there is at equilibrium about twice as much trans-butene as cis-butene in the gas mixture, that is, $[P_B/P_A]^{eq} \approx 2$, where superscript "eq" means equilibrium value. This implies that if we start from a mixture containing a lower amount of trans-butene, the reaction will go to the right until the cis-butene content has decreased and trans-butene content has increased so much that the equilibrium value is reached. If, on the other hand, the amount of trans-butene is higher, the reaction will instead go to the left. We will now examine why this is the case and we start by determining how Gibbs energy varies for different compositions of the mixture when the total pressure $P = P_A + P_B$ is constant.

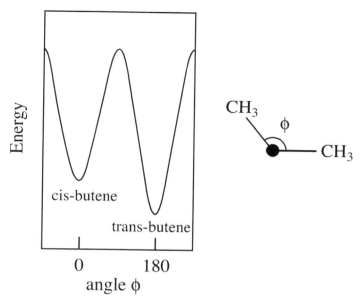

Figure 5.3   Schematic diagram of the energy as a function of the rotation angle for the methyl groups around the double bond in 2-butene. The illustration on the right shows how the angle is measured when one sees the molecule along the double bond (marked as a filled circle).

In Figure 5.4 we have plotted G from Equation (5.7) for one mole of a mixture of cis-butene and trans-butene as a function of the partial pressure. (Since we have one mole, the notation G_m is used in the figure.) Values for G_m^0(cis-butene) and G_m^0(trans-butene) have been taken from published data. We have chosen the case with a total pressure $P_A + P_B = 1$ bar, that is, the total pressure is equal to P^0. The curve has a minimum located at $P_B = 0.68$ bar and $P_A = 0.32$ bar, i.e., when $P_B/P_A = 2.1$. To the left of the minimum point, the conversion of A to B is spontaneous (G_m decreases when going to the right, i.e., when P_B increases). At the minimum point the system is at equilibrium (compare with Figure 4.15). To the right of this point the reverse reaction is spontaneous (G_m increases when going to the right, but decreases in the opposite direction).

Let us now investigate in more detail how G changes for the reaction at constant P.

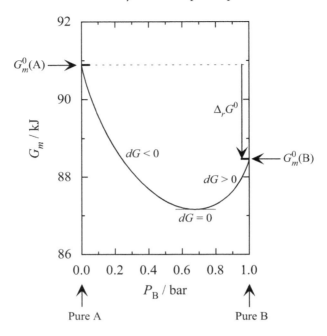

Figure 5.4   Gibbs energy G_m for one mole of a mixture of cis-butene (A) and trans-butene (B) plotted as function of the partial pressure trans-butene P_B. The total pressure is 1 bar so the partial pressure cis-butene is $P_A = 1 - P_B$ bar, which means that P_A is 1 bar to the left on scale and 0 to the right. The temperature is 400 K. Gibbs energy $G_m^0(A)$ and $G_m^0(B)$ for pure A and B at pressure $P^0 = 1$ bar are marked by thick horizontal lines (the values of these two quantities are taken from literature data). $\Delta_r G^0$ is the difference between $G_m^0(B)$ and $G_m^0(A)$, and is illustrated by the downward arrow. When G_m as a function of P_B decreases we have $dG < 0$, and when it increases $dG > 0$. At the minimum point $dG = 0$.

A LITTLE DERIVATION

We assume that we have a mixture with partial pressures P_A and P_B. Gibbs energy of the mixture is given by Equation (5.7) which we can write as

$$G = \frac{N_A}{N_{Av}}\left[G_m^0(A) + RT\ln\left(\frac{P_A}{P^0}\right)\right] + \frac{N_B}{N_{Av}}\left[G_m^0(B) + RT\ln\left(\frac{P_B}{P^0}\right)\right], \quad (5.9)$$

where N_A and N_B are the number of molecules in the mixture (we have divided by Avogadro's constant N_{Av} to obtain the number of moles of A and B).

If an A molecule is converted to a B molecule, we obtain $N_A - 1$ molecules of species A and $N_B + 1$ molecules of B. From Equation (5.9) we see that Gibbs energy then becomes[5]

$$\frac{N_A - 1}{N_{Av}}\left[G_m^0(A) + RT\ln\left(\frac{P_A}{P^0}\right)\right] + \frac{N_B + 1}{N_{Av}}\left[G_m^0(B) + RT\ln\left(\frac{P_B}{P^0}\right)\right].$$

Thereby G has been changed by dG, which is the difference between this expression and Equation (5.9), that is,

$$dG = \frac{-1}{N_{Av}}\left[G_m^0(A) + RT\ln\left(\frac{P_A}{P^0}\right)\right] + \frac{1}{N_{Av}}\left[G_m^0(B) + RT\ln\left(\frac{P_B}{P^0}\right)\right].$$

If we introduce $dn = 1/N_{Av}$, we can write this as

$$dG = \left[G_m^0(B) - G_m^0(A) + RT\ln\left(\frac{P_B}{P^0}\right) - RT\ln\left(\frac{P_A}{P^0}\right)\right]dn. \qquad (5.10)$$

Thus, dG is the small change in G when the number of moles of A changes by $dn_A = -dn$ and of B by $dn_B = dn$, corresponding to one molecule each of A and B.[6]

Equation (5.10) gives the change in Gibbs energy when an A molecule is converted to a B molecule in a mixture with the partial pressures P_A and P_B. Since

$$\ln\left(P_B/P^0\right) - \ln\left(P_A/P^0\right) = \ln\left(\frac{P_B/P^0}{P_A/P^0}\right) = \ln\left(\frac{P_B}{P_A}\right)$$

we can write Equation (5.10) as

$$dG = \left[\Delta_r G^0 + RT\ln\left(\frac{P_B}{P_A}\right)\right]dn \qquad (5.11)$$

where

$$\Delta_r G^0 = G_m^0(B) - G_m^0(A) \qquad (5.12)$$

---

[5]One can show that there is no contribution from changes in partial pressures (this is proven in footnote 6).

[6]The expression (5.10) actually applies generally when dn moles react according to the formula A $\rightarrow$ B, provided dn is small. Thereby dG is the change in G for dn moles. (Mathematically, one can derive this result by obtaining the differential of G from Equation (5.7) and inserting $dn_A = -dn$ and $dn_B = dn$. Thereby one uses that $n_A RT d\ln P_A + n_B RT d\ln P_B = n_A RT dP_A/P_A + n_B RT dP_B/P_B = V dP_A + V dP_B = V dP$, where V is the volume of the vessel that contains the A and B gases and we have used the ideal gas law. $V dP$ is zero since P is constant, so the small changes in partial pressures give no contribution.)

is the difference in Gibbs energy between one mole pure B and one mole pure A at temperature T and pressure P^0 (the subscript r in $\Delta_r G^0$ stands for "reaction").[7] In Figure 5.4 we have indicated $\Delta_r G^0$ by a large vertical arrow to the right. We have defined dn in Equation (5.11) as a positive number when A is converted to B. Therefore, the sign of dG is determined by the expression in square brackets in this equation. A negative dG (decreasing G) means that the conversion of A to B is spontaneous and this therefore occurs when $\Delta_r G^0 + RT \ln(P_B/P_A) < 0$. If the expression is positive we have $dG > 0$, and then, instead, the reverse reaction is spontaneous, i.e., the conversion of B to A.[8]

As we saw in Figure 5.4, to the left of the minimum point, the conversion of A to B is spontaneous ($dG < 0$) and to the right of it the reverse reaction is spontaneous ($dG > 0$). At the minimum point we have $dG = 0$ and then there is equilibrium (like the minimum in Figure 4.15). This means, according to Equation (5.11), that we have equilibrium when the expression in square brackets is zero. The equilibrium condition is therefore

$$\Delta_r G^0 + RT \ln\left(\left[\frac{P_B}{P_A}\right]^{eq}\right) = 0. \tag{5.13}$$

As we have seen, this occurs when $[P_B/P_A]^{eq} = 2.1$ in our example.

We can write Equation (5.13) as $\ln[P_B/P_A]^{eq} = -\Delta_r G^0/RT$. This means that $[P_B/P_A]^{eq}$ at a given temperature is equal to a number which is determined by $\Delta_r G^0$. In our example in Figure 5.4, we assumed that the total pressure $P_A + P_B$ for the mixture is 1 bar, but Equation (5.13) does not require this. The total pressure can be anything. At equilibrium, the ratio between P_B and P_A still has the same value. If, for example, $P_A + P_B$ is 0.5 bar, we still have $[P_B/P_A]^{eq} = 2.1$ at equilibrium and thus $P_B = 0.34$ bar and $P_A = 0.16$ bar. This means that no matter how much A and B we have in our ideal gas mixture from the beginning and regardless of the total pressure, equilibrium is reached when the ratio P_B/P_A becomes equal to a certain number that

[7]The symbol Δ_r symbolizes change of some quantity during a reaction according to the stoichiometry of the reaction formula (in moles).

[8]Here, we have chosen to look at the change of G when A is converted to B. This does not mean that the transformation of molecules takes place only in one direction. When, for example, the reaction A $\rightarrow$ B is spontaneous, more A-molecules are converted to B-molecules per unit time than B-molecules are converted to A-molecules. At equilibrium, an equal number is converted on average in either direction.

is the same in all these cases (provided the temperature is the same). This number is denoted K and is called the *equilibrium constant*

$$\left[\frac{P_B}{P_A}\right]^{eq} = K. \tag{5.14}$$

The relationship (5.14) is called the **law of mass action** as applied to the reaction $A \rightleftharpoons B$. In our example $K = 2.1$.

Since $\ln[P_B/P_A]^{eq} = -\Delta_r G^0/RT$ it follows that

$$\ln K = -\frac{\Delta_r G^0}{RT}, \tag{5.15}$$

which implies that $K = \exp(-\Delta_r G^0/RT)$. Hence, the equilibrium constant for a mixture at a given temperature T is determined by $\Delta_r G^0$, which in turn is determined by the *properties of the pure substances* at pressure P^0 and temperature T.

Let us now do a molecular interpretation of our results. First we consider the influence of the configurational entropy, S_{conf}, on the curve in Figure 5.4, which shows G for the mixture as obtained from Equation (5.7). In fact, the configurational entropy of the mixture of A and B is as large as possible when the system contains *equal* amounts of A and B. This can be understood from the following argument.[9]

THE INFLUENCE OF CONFIGURATIONAL ENTROPY ALONE

The logarithmic terms in Equation (5.7) and hence the logarithm in Equation (5.11) originates from S_{conf} (this follows from the derivation of Equations (5.3) and (5.7)), If we would have $G_m^0(A) = G_m^0(B)$, we see from Equation (5.12) that $\Delta_r G^0 = 0$. When $\Delta_r G^0 = 0$, only the configurational entropy contributes to the equilibrium condition and from Equation (5.15) follows that $K = 1$, so equilibrium would occur when the system contains equal amounts of A and B, i.e., when $P_A = P_B$. Thus, if S_{conf} alone would determine the equilibrium, it would

---

[9]In order to realize this by instead using an argument where one counts the number of configurations (as we did in Sections 2.3–2.5), one must take into account that particles of the same kind (for example, of species A) cannot be distinguished from each other (this we did *not* do in Section 2.3 as noted in footnote 13 there). If we do this, the constant $\mathcal{K}$ in Equation (2.2) becomes $\mathcal{K} = 1/(N!\nu^N)$, where the factor $N!$ asserts that permutations between particles of the same kind (for example, that molecules 1 and 2 change places with each other) are counted as the same configuration. The argument presented here avoids this complication, but it is still correct because it is based on Equation (2.3) where $\mathcal{K}$ is eliminated.

occur at $P_B = 0.5$ and the minimum for G in Figure 5.4 would accordingly be located there (the curve would then be symmetrical around this point).[10] Remember that entropy contributes with a term $-TS$ to G, so a minimum in G corresponds in the present case to a maximum in S, so S_{conf} is as as large as possible when $P_A = P_B = 0.5$ bar. The configurational entropy accordingly favors the reaction where the component with the highest partial pressure is converted to that with the lowest. When $P_A > P_B$ (that is, $P_B < 0.5$ bar in the present case), the configurational entropy increases when A $\rightarrow$ B (which is favorable), while for $P_A < P_B$ (i.e., $P_B > 0.5$ bar) there is a decrease in S_{conf} when A $\rightarrow$ B (which is unfavorable). In the latter case, if the reverse process would occur (B $\rightarrow$ A), S_{conf} would increase instead. After this investigation of S_{conf}, let us now return to our case where $G_m^0(A) \neq G_m^0(B)$.

The reason why we do not obtain a mixture with $P_A = P_B$ at equilibrium is that apart from a change in configurational entropy there is another contribution when A is converted to B and vice versa. In our example where A = cis-butene and B = trans-butene, we have seen that cis-butene has a higher energy than trans-butene, which means that energy is released when A is converted to B. The released energy is distributed throughout the environment, which gives a positive contribution to the total entropy (a negative contribution to G). This means that the total entropy can increase when A is converted to B even when $P_B > 0.5$ bar ($P_A < P_B$), despite that the configuration entropy then decreases. The condition for this to be the case is that the increase in entropy caused by the energy release is larger than the decrease in S_{conf}.

The release of energy therefore shifts the equilibrium so much that there is more B than A in the mixture, but not so much that the amount of B becomes too large. If P_B becomes too large compared to P_A, the decrease in S_{conf} will be so great that the total entropy does not increase when A is converted to B. This is because the larger the difference between P_B and P_A, the greater the change in S_{conf} during the conversion, which limits how far the equilibrium is shifted. As we

---

[10]This symmetry can also be understood from the fact that the sum of the two logarithmic terms in Equation (5.8) would be unchanged if subscripts A and B were swapped. When $G_m^0(A) = G_m^0(B)$ the same applies to the entire right-hand side in the equation as well as to the condition $n_A + n_B =$ constant that is used in the current case.

saw earlier, at $T = 400$ K, there is about twice as much trans-butene (B) than cis-butene (A) in the gas mixture at equilibrium.

The energy change when A (cis-butene) is converted to B (trans-butene) is included in the term $\Delta_r G^0$ in Equation (5.11). The quantity $\Delta_r G^0$, which is given by Equation (5.12), is equal to the change in Gibbs energy when one takes one mole pure A at pressure P^0 and converts it *completely* to one mole pure B at the same pressure (see the downward arrow labeled $\Delta_r G^0$ in Figure 5.4). The energy change during the transformation A $\rightarrow$ B contributes to $\Delta_r G^0$ since $G = U - TS + PV$ and therefore $\Delta G = \Delta U - T\Delta S + \Delta(PV)$. The most significant contribution to $\Delta_r G^0$ in this case is the energy change. There is also an entropy contribution in $\Delta_r G^0$, which we shall discuss shortly. Since the total number of moles of gas does not change during the reaction, the volume does not change so $\Delta(PV) = P\Delta V = 0.$[11]

Since the gases are ideal, we have no interaction between the molecules. The change in energy when we convert an A molecule to a B molecule is thus independent of the presence or absence of other molecules. Therefore, the energy change per converted molecule is the same at pressure P^0 as at the actual partial pressures in the mixture. Furthermore, cis-butene and trans-butene molecules have different entropies because they have different molecular structures,[12] so there is an entropy contribution in $\Delta_r G^0$. This entropy difference gives a smaller contribution to $\Delta_r G^0$ than the difference in energy (at least if the temperature is not very high), so the energy contribution dominates for the conversion between cis-butene and trans-butene.[13] The entropy contribution is, however, not negligible. At 400 K, it is

---

[11]Since $PV = nRT$ for an ideal gas, we have $\Delta(PV) = RT\Delta n$, where Δn is the change in total number of moles and T is constant. When $\Delta n = 0$ it is clear that $\Delta(PV) = 0$. Furthermore, this means that there is no difference between the changes in enthalpy and energy in this case. We have $H = U + PV$ and therefore $\Delta H = \Delta U + RT\Delta n$. When $\Delta n = 0$ we have $\Delta H = \Delta U$ for an ideal gas at constant T.

[12]The entropy difference is due to the fact that the energy of a cis-butene molecule can be distributed between the internal vibrations, rotations, and electronic states in different ways than the energy of a trans-butene molecule. (A difference in number of possible molecular conformations can in the general case also give an entropy difference.)

[13]In the general case, $\Delta_r G^0$ also contains a contribution from the configurational entropy of the gases. Here, however, there is no such contribution because this case concerns the conversion between two isomers and the pressure is equal to P^0, both before and after the conversion.

about half as large as the energy contribution. Consequently, it too affects the location of the equilibrium point.

5.2.2   Equilibrium of type $A \rightleftharpoons 2B$

Now let us examine a case where the number of moles of gas changes during the reaction, which is the most common type. We take the simplest example of such a reaction, namely, when a molecule is dissociated into two identical molecules

$A \rightleftharpoons 2B,$

for instance $A = N_2O_4$ and $B = NO_2$ in gas phase

$N_2O_4(g) \rightleftharpoons 2NO_2(g).$

At 25°C and 1 bar about 20% of N_2O_4 is dissociated into NO_2 at equilibrium. If we start from a mixture with excess of A and let the reaction take place until equilibrium is reached, the system volume must be increased for the pressure to remain constant because two B molecules are formed per A molecule. We will now determine how much Gibbs energy changes during the reaction. The reasoning is very similar to what we did when we came up with Equation (5.10).

A LITTLE DERIVATION

We assume that we have a mixture with partial pressures P_A and P_B that contains N_A molecules of A and N_B of B. Gibbs energy of the mixture is still given by Equation (5.9). When one A molecule is converted into two B molecules, we obtain $N_A - 1$ molecules of species A and $N_B + 2$ molecules of B. From Equation (5.9) we see that Gibbs energy then becomes

$$\frac{N_A - 1}{N_{Av}}\left[G_m^0(A) + RT\ln\left(\frac{P_A}{P^0}\right)\right] + \frac{N_B + 2}{N_{Av}}\left[G_m^0(B) + RT\ln\left(\frac{P_B}{P^0}\right)\right],$$

which implies that G has changed by

$$dG = \frac{-1}{N_{Av}}\left[G_m^0(A) + RT\ln\left(\frac{P_A}{P^0}\right)\right] + \frac{2}{N_{Av}}\left[G_m^0(B) + RT\ln\left(\frac{P_B}{P^0}\right)\right].$$

If we again introduce $dn = 1/N_{Av}$, we can write this as (compare with Equation (5.10))

$$dG = \left[2G_m^0(B) - G_m^0(A) + 2RT\ln\left(\frac{P_B}{P^0}\right) - RT\ln\left(\frac{P_A}{P^0}\right)\right]dn. \qquad (5.16)$$

Thus, dG is the small change in G when the number of moles of A changes by $dn_A = -dn$ and of B by $dn_B = 2dn$, corresponding to one molecule of A and two of B.[14]

Equation (5.16) gives the change in Gibbs energy when an A molecule is converted to two B molecules in a mixture with partial pressures P_A and P_B. Now, since

$$2\ln\left(\frac{P_B}{P^0}\right) - \ln\left(\frac{P_A}{P^0}\right) = \ln\left(\left\{\frac{P_B}{P^0}\right\}^2\right) - \ln\left(\frac{P_A}{P^0}\right)$$

$$= \ln\left(\frac{\left\{\frac{P_B}{P^0}\right\}^2}{\frac{P_A}{P^0}}\right)$$

we can write Equation (5.16) as

$$dG = \left[\Delta_r G^0 + RT\ln\left(\frac{\left\{\frac{P_B}{P^0}\right\}^2}{\frac{P_A}{P^0}}\right)\right]dn \tag{5.17}$$

where

$$\Delta_r G^0 = 2G_m^0(B) - G_m^0(A) \tag{5.18}$$

is the change in Gibbs energy at temperature T when one takes one mole pure A at pressure P^0 and converts it *completely* to two moles pure B at the same pressure (thus one starts with the reactant in its standard state and finishes with the product in its standard state). Note that $\Delta_r G^0$ can be calculated directly from known (tabulated) values for the pure substances.

The most significant difference from the previous case is that P_B/P^0 is raised to the power 2. This is a consequence of the fact that we have a coefficient 2 in front of B in the reaction formula $A \rightleftharpoons 2B$, as apparent from the preceding derivation. Otherwise, all is much the same. In Figure 5.5, G from Equation (5.7) is plotted for a mixture of N_2O_4 and NO_2 (for a total amount equivalent to one mole of

---

[14]The expression (5.16) applies generally when dn moles react according to the formula $A \rightarrow 2B$, provided dn is small (compare with footnote 6). Thereby dG is the change in G when dn moles A are converted to $2dn$ moles B.

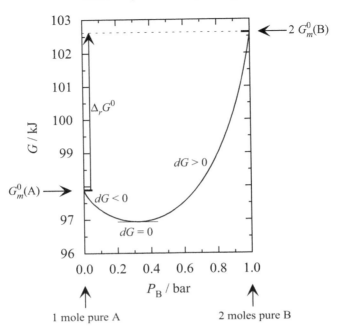

Figure 5.5   Gibbs energy G for a mixture of N_2O_4 (A) and NO_2 (B) plotted as a function of the partial pressure of NO_2, P_B. The total pressure P_A+P_B is 1 bar and the temperature is 298 K. At the left end of the pressure range, we have one mole of pure A and at the right end two moles of pure B. For intermediate partial pressures, we have a mixture where N_2O_4 has been dissociated into NO_2 to varying extents. Gibbs energy at the end points of the curve, $G_m^0(A)$ and $2G_m^0(B)$, respectively, are marked by thick horizontal lines (values from literature data). $\Delta_r G^0$ is the difference between these two values.

pure N_2O_4 that is partly dissociated). The total pressure is constant, $P_A + P_B = 1$ bar. The analogy to Figure 5.4 is evident.

It is practical to introduce the notation

$$Q = \frac{\left\{\frac{P_B}{P^0}\right\}^2}{\frac{P_A}{P^0}}, \tag{5.19}$$

which is called the **reaction quotient** for the reaction $A \rightleftharpoons 2B$, whereby Equation (5.17) can be written

$$dG = \left[\Delta_r G^0 + RT \ln Q\right] dn. \tag{5.20}$$

Note that the value of Q depends on the partial pressures P_A and P_B of the mixture.[15]

We derived Equation (5.20) for the conversion of one molecule A to two B, but it is easy to show that it applies generally to the change in G when a small amount dn moles is converted according to the reaction formula A $\rightarrow$ 2B. The equation can alternatively be written[16]

$$\Delta_r G = \Delta_r G^0 + RT \ln Q, \qquad (5.21)$$

where $\Delta_r G$ is the change in G for the conversion of *one mole* according to the reaction formula when there is *a very large amount of reactants and/or products in the reaction mixture.*

In Equation (5.19), the reaction quotient Q is expressed in terms of partial pressures. One can also express it in terms of the molecular densities ρ_A and ρ_B (the number of molecules per unit volume). According to the ideal gas law the pressure is proportional to the density (when T is constant), and we can use Equation (5.2) to write Equation (5.19) in the form

$$Q = \frac{\left\{\frac{\rho_B}{\rho^0}\right\}^2}{\frac{\rho_A}{\rho^0}}, \qquad (5.22)$$

where ρ^0 is the density at pressure P^0. Alternatively, one can use the concentration c as a measure of content. Since $\rho_i = c_i N_{Av}$, Equation (5.22) yields

$$Q = \frac{\left\{\frac{c_B}{c^0}\right\}^2}{\frac{c_A}{c^0}}, \qquad (5.23)$$

where $c^0 = \rho^0/N_{Av} = P^0/RT$.

The sign of dG in Equation (5.20) is determined by the expression within the square brackets. When the expression is negative,

---

[15]Analogously, one can define the reaction quotient in Equation (5.11) as P_B/P_A or rather as $\{P_B/P^0\}/\{P_A/P^0\}$, which is the same thing. By making the corresponding definition of Q in other cases, Equation (5.20) becomes valid for the general case.

[16]The condition for Equation (5.20) is that the quantity dn that reacts is small. If we have a system with a huge amount of A and B, a conversion of *one mole* is so small that the partial pressures P_A and P_B, and hence Q, will not be changed significantly. Equation (5.20) can then be written as Equation (5.21), where we have replaced dG with $\Delta_r G$ and dn with 1. This form of the equation is commonly found in textbooks, but the conditions under which it applies are not always clearly stated.

$dG < 0$ and reaction A $\rightarrow$ 2B is spontaneous. If instead the expression is positive, $dG > 0$ and the reverse reaction 2B $\rightarrow$ A is spontaneous. By inserting the partial pressures P_A and P_B for a mixture with arbitrary composition in Equations (5.19) and (5.20), one can hence determine whether the reaction goes spontaneously in one direction or the other. In this manner one can find out whether the spontaneous process for that particular composition is that more A molecules are dissociated than B molecules are merged or the opposite. A high P_A and low P_B favors the former, while a low P_A and high P_B favors the latter. Since P_B is squared in Equation (5.19), a changed amount of B can very easily affect this balance. Likewise ρ_B is squared in Equation (5.22).[17]

Just as in the former case we have equilibrium when $dG = 0$, that is, when the square bracket in Equation (5.20) is zero. The condition for equilibrium is therefore

$$\Delta_r G^0 + RT \ln Q^{eq} = 0, \tag{5.24}$$

where Q^{eq} is the value of Q at equilibrium. We can write Equation (5.24) as $\ln Q^{eq} = -\Delta_r G^0 / RT$, which implies that equilibrium is attained at a certain temperature when Q is equal to a number that is determined by $\Delta_r G^0$. This number is what we call K, the equilibrium constant, and we see that in this case too it is determined by $\ln K = -\Delta_r G^0 / RT$ (Equation (5.15)). No matter how much of the substances A and B we have in our ideal gas mixture from the beginning and regardless of the total pressure, equilibrium is reached when Q becomes equal to the number K in all cases. Thus we have the equilibrium condition $Q^{eq} = K$, which according to Equation (5.19) implies

---

[17]The fact that the square of ρ_B is essential for the direction in which the reaction goes can also be understood from the following arguments. The probability to find a B molecule in a small volume element is proportional to the density ρ_B. The probability to find another B molecule in the same volume element is proportional to ρ_B too. This means that the probability that the two molecules will be there at the same time is proportional to the product of these probabilities, that is, $(\rho_B)^2$. The number of B molecules which merge per unit time in the system is thus proportional to $(\rho_B)^2$, because two such molecules must be in the same place at the same time in order to react. The number of A molecules that dissociates per time unit is, on the other hand, proportional to the number of such molecules in the system, which is proportional to the ρ_A. The ratio $(\rho_B)^2 / \rho_A$ is therefore of importance to determine whether a larger number of A molecules dissociate, than B molecules merge per unit time, or vice versa. This is consistent with Equation ((5.22)), where $(\rho_B)^2 / \rho_A$ appears.

that

$$\left[\frac{\left\{ \frac{P_B}{P^0} \right\}^2}{\frac{P_A}{P^0}} \right]^{eq} = K. \tag{5.25}$$

For partial pressures that fulfill this condition, G is minimal and then an equal number of A molecules dissociate per unit of time as A is being formed by the merger of B molecules. Equation (5.25) is the equilibrium expression for the reaction A $\rightarrow$ 2B, giving the law of mass action for this case. We can alternatively write it as

$$\left[\frac{(P_B)^2}{P_A} \right]^{eq} \frac{P^0}{(P^0)^2} = K,$$

that is,

$$\left[\frac{(P_B)^2}{P_A} \right]^{eq} = K_P, \tag{5.26}$$

where $K_P = KP^0$. While K is a dimensionless number, K_P has the same unit as that used for the pressures P^0, P_A, and P_B. Thus, K_P is an *equilibrium constant in pressure units*.

For the example in Figure 5.5, where the total pressure is 1 bar and $T = 298$ K, G has minimum when $P_B = 0.32$ and $P_A = 0.68$ bar. The value of the equilibrium constant $K_P = 0.148$ bar $= 1.48 \cdot 10^4$ Pa. The first value should be used in Equation (5.26) when the partial pressures have the unit bar and the second when they have the unit Pa. The value of K in Equation (5.25) is equal to the dimensionless number 0.148 regardless of which unit one uses for the pressures. This is one of the reasons why it is useful to write the equilibrium condition as Equation (5.25) rather than (5.26). In calculations it is, however, often more practical to use (5.26).

Say that we would like to find out the equilibrium values of the partial pressures for a gas mixture of N_2O_4 (A) and NO_2 (B) when the total pressure is, for example, 0.5 bar and $T = 298$ K. We then use Equation (5.26) with the same value of the equilibrium constant, $K_P = 0.148$ bar. We insert $P_A = 0.5 - P_B$ in Equation (5.26) and obtain $(P_B)^2/(0.5 - P_B) = 0.148$, where we specify the pressures in bar. The solution to this equation (which can be written as a quadratic equation) is $P_B = 0.21$ bar and we obtain $P_A = 0.5 - P_B = 0.29$ bar.

If we compare to the previous case where the total pressure was 1 bar and $P_B = 0.32$ bar, we see that the partial pressure $P_B = 0.21$ bar is a larger proportion of the total pressure 0.5 bar in this latter case. Thus, we obtain an increased percentage NO_2 when we lower the total pressure. The amount of N_2O_4 that is dissociated at equilibrium, accordingly increases with decreasing pressure (we say that the degree of dissociation increases). At 1 bar the degree of dissociation is 19% and at 0.5 bar it is 26%, that is, if we had pure N_2O_4 from the beginning, then, at equilibrium, 19% and 26% of N_2O_4 has dissociated, respectively. The growing degree of dissociation when one reduces the total pressure by increasing the volume can be understood molecularly in the following manner. When the volume is increased the system gains configurational entropy by dissociating N_2O_4 and thereby forming a larger number of molecules which spread in the volume. How much the total entropy changes (and thus how much is actually dissociated) depends also on the enthalpy and entropy changes for the dissociation itself. This is taken into account in the equilibrium condition via $\Delta_r G^0$ and therefore affects the value of the equilibrium constant. The shift in degree of dissociation of N_2O_4 when the pressure is decreased is an example of what is called *Le Châtelier's principle*,[18] namely, that an equilibrium that is disturbed will be shifted in the direction that counteracts the cause of the disturbance.

The law of mass action can also be expressed in terms of concentrations. If we insert Equation (5.23) in the equilibrium condition $Q^{eq} = K$ we obtain

$$\left[\frac{\left\{ \frac{c_B}{c^0} \right\}^2}{\frac{c_A}{c^0}} \right]^{eq} = K \tag{5.27}$$

and we can alternatively write it as

$$\left[\frac{(c_B)^2}{c_A} \right]^{eq} = K_c, \tag{5.28}$$

where $K_c = K c^0 = K P^0/RT$. Hence K_c, the *equilibrium constant in concentration units*, has the same unit as that used for the concentrations c^0, c_A, and c_B, for example, M (molar = moles per liter) or the SI unit $mol\, m^{-3}$. In our example with N_2O_4 and NO_2 at 298 K, the equilibrium constant K in Equation (5.27) is, of course, still 0.148 and dimensionless, while $K_c = 0.148\, P^0/RT = 6.0\, mol\, m^{-3} = 6.0 \cdot 10^{-3}\, M$ should be

---

[18]Henry-Louis Le Châtelier (1850–1936) was a French chemist who is most known for his work regarding this principle.

used in Equation (5.28). K is called the **thermodynamic equilibrium constant** in order to distinguish it from the other equilibrium constants K_P and K_c, which have units and have values that depend on the units one uses.

5.2.3 Equilibria of type aA+bB ⇌ xX+yY

The reasoning above is easily generalized to arbitrary reactions in ideal gas mixtures and we will only give the main results without derivations. For example, the reaction

$$aA + bB \rightleftharpoons xX + yY$$

with stoichiometric coefficients a, b, x, and y, has the reaction quotient

$$Q = \frac{\left\{\frac{P_X}{P^0}\right\}^x \left\{\frac{P_Y}{P^0}\right\}^y}{\left\{\frac{P_A}{P^0}\right\}^a \left\{\frac{P_B}{P^0}\right\}^b}, \tag{5.29}$$

where P_A, P_B, P_X, and P_Y are the partial pressures of substances A, B, X, and Y in the gas mixture one is dealing with. For a A-molecules that react with b B-molecules and form x X-molecules and y Y-molecules in the mixture, the change dG in Gibbs energy is given by Equation (5.20) with Q taken from Equation (5.29). In this case, $\Delta_r G^0$ is the change in Gibbs energy when a moles of pure A and b moles of pure B reacts completely to form x moles of pure X and y moles of pure Y, where the pressure of all pure gases is equal to P^0, as illustrated in Figure 5.6. We have

$$\begin{aligned} \Delta_r G^0 &= G^0_{\text{after}} - G^0_{\text{before}} = G^0_{\text{products}} - G^0_{\text{reactants}} \\ &= xG^0_m(X) + yG^0_m(Y) - [aG^0_m(A) + bG^0_m(B)]. \end{aligned} \tag{5.30}$$

The equilibrium condition is still $dG = 0$, which implies, as before, that we have $Q^{\text{eq}} = K$, where K is the equilibrium constant given by Equation (5.15) that implies

$$K = e^{-\frac{\Delta_r G^0}{RT}}. \tag{5.31}$$

The equilibrium expression is therefore

$$\left[\frac{\left\{\frac{P_X}{P^0}\right\}^x \left\{\frac{P_Y}{P^0}\right\}^y}{\left\{\frac{P_A}{P^0}\right\}^a \left\{\frac{P_B}{P^0}\right\}^b}\right]^{\text{eq}} = K \tag{5.32}$$

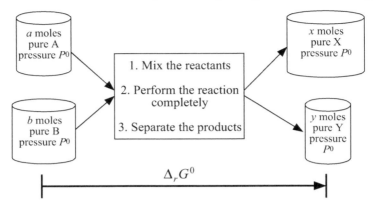

Figure 5.6 $\Delta_r G^0$ is the total change in G for the following process: The re-actants, a moles of pure A and b moles of pure B at pressure P^0, are passed into a reaction chamber (whereby their pressures change) and are made to react completely to form the products, x moles of X and y moles of Y, which are separated from the reaction mixture and given pressure P^0. Note that G in the general case is changed not only during the reaction but also during mixing and separation.

in this case and gives the law of mass action for this more general case. The equilibrium expressions in the previous subsections are special cases of this more general equation.

We can alternatively write Equation (5.32) as

$$\left[\frac{(P_X)^x (P_Y)^y}{(P_A)^a (P_B)^b} \right]^{eq} = K_P, \tag{5.33}$$

where $K_P = K(P^0)^{\Delta N_r}$ and where ΔN_r is the change in number of molecules during the reaction according to reaction formula, namely,

$$\Delta N_r = x + y - a - b.$$

Note that K_P therefore has the unit $(bar)^{\Delta N_r}$ when the pressures are measured in bars. One can also express the equilibrium condition in terms of concentrations similar to Equations (5.27) and (5.28).

Key points

- One can determine whether a chemical reaction can take place or not for an arbitrary mixture of reactants and products by examining Gibbs energy for the mixture (T and P are assumed to be constant). If G decreases as the reactants form products at the actual composition, the reaction can occur spontaneously. If G increases, the reaction cannot take place, but instead the reverse reaction can occur spontaneously. Equilibrium is reached when the composition is such that G assumes its minimal value.

• These criteria can be expressed in a rational manner by using two quantities, $\Delta_r G^0$ and Q, which are defined as:

$\Delta_r G^0$ is the change in Gibbs energy when one starts from pure reactants at pressure P^0, mixes the reactants, performs the reaction *completely*, and finally has clean, separated products at pressure P^0.

Q, called the reaction quotient, is determined by the composition of the mixture and by the reaction formula.

• For the reaction $a A + b B \rightleftharpoons x X + y Y$ in the gas phase (ideal gases) the reaction quotient is

$$Q = \frac{\left\{\frac{P_X}{P^0}\right\}^x \left\{\frac{P_Y}{P^0}\right\}^y}{\left\{\frac{P_A}{P^0}\right\}^a \left\{\frac{P_B}{P^0}\right\}^b},$$

where P_A, P_B, P_X, and P_Y are the partial pressures in the gas mixture one is dealing with.

• If $\Delta_r G^0 + RT \ln Q$ is negative, the reaction can occur spontaneously to the right. If the expression is positive, the reaction can happen spontaneously to the left, and if it is zero, one has chemical equilibrium.

• The equilibrium condition can also be expressed by the expression $Q^{eq} = K$ where K is the thermodynamic equilibrium constant, which can be calculated from $\Delta_r G^0$ according to $K = \exp(-\Delta_r G^0 / RT)$.

• For the reaction $a A + b B \rightleftharpoons x X + y Y$ one has the equilibrium expression (law of mass action)

$$\left[\frac{\left\{\frac{P_X}{P^0}\right\}^x \left\{\frac{P_Y}{P^0}\right\}^y}{\left\{\frac{P_A}{P^0}\right\}^a \left\{\frac{P_B}{P^0}\right\}^b}\right]^{eq} = K.$$

It can also be written

$$\left[\frac{(P_X)^x (P_Y)^y}{(P_A)^a (P_B)^b} \right]^{\mathrm{eq}} = K_P,$$

where $K_P = K(P^0)^{\Delta N_r}$ with $\Delta N_r = x + y - a - b$ or

$$\left[\frac{(c_X)^x (c_Y)^y}{(c_A)^a (c_B)^b} \right]^{\mathrm{eq}} = K_c,$$

where $K_c = K(c^0)^{\Delta N_r}$. While K is unitless, K_P and K_c have units.

CHAPTER 6

Phases and temperature variations

In the previous chapter we saw how we can utilize the thermodynamical quantities and relationships obtained earlier to understand chemical reactions and equilibria. The general nature of thermodynamical concepts and reasoning is further illustrated in the present chapter, where we will investigate other applications of thermodynamics. First, we shall investigate how matter changes its state of aggregation between gas, liquid, and solid under various conditions. Thereby we will, for example, look in more detail into the vaporization and condensation processes that we investigated earlier. In particular we will focus on Helmholtz and Gibbs energies to see how they can be utilized to understand what phases are present under various conditions and the transitions between them. Thus we will familiarize ourselves with how to use the thermodynamic "machinery" that we have built up in order to see how and why thermodynamic arguments work for these kinds of questions. Finally, in Section 6.2 we shall investigate how various thermodynamical quantities depend on temperature, building on what we already know about this.

6.1   To boil and to freeze
Phase transitions

When we investigated the evaporation of a liquid droplet in Section 3.2, we saw that the balance between entropy and energy is crucial for what happens. For molecules to be able to leave the liquid phase, energy is required, which is taken from the environment. This gives a negative contribution to the total entropy because the number of possibilities to distribute the energy then decreases. At the same time, the molecules get access to a larger number of configurations, which gives a positive entropy contribution. If the latter contribution dominates, then the liquid evaporates spontaneously, and if the former dominates, then the vapor condenses spontaneously. In Section 3.8, we saw that this balance can be expressed in a practical manner by

using the concept of free energy and we will begin by recapitulating
the results there.

We assume that the liquid and its vapor are in a closed vessel with
constant volume V and we keep the temperature T constant. The contents of the vessel is our system and we assume that there is only one
species present, so the liquid and gas are pure. Let ΔU be the change
in energy and ΔS the change in entropy of the system when a quantity
of liquid is vaporized, i.e., for the process fluid $\rightarrow$ vapor. The change
in Helmholtz energy is $\Delta A = \Delta U - T\Delta S$ and the change in total entropy of the system and the surroundings is according to Equation
(3.8) given by

$$\Delta S_{\text{tot}} = -\frac{\Delta A}{T} = \Delta S - \frac{\Delta U}{T}, \quad (6.1)$$

where $-\Delta U/T$ is the entropy change of the surroundings. We have
$\Delta U > 0$ because the molecules in the vapor have higher energy than
in the liquid and $\Delta S > 0$ because the entropy of the vapor is higher.
When $\Delta A < 0$ ($\Delta S_{\text{tot}} > 0$) evaporation is spontaneous and when $\Delta A > 0$
($\Delta S_{\text{tot}} < 0$) condensation is spontaneous, i.e., the reverse process (compare with Figure 3.18). We have equilibrium when A does not change
upon evaporation of a small amount of liquid (or condensation of a
small amount of vapor), $dA = 0$, and then the pressure of the vapor is
equal to what we call the vapor pressure of the liquid at the temperature in question.

What happens if we change the temperature but keep the volume
constant? From Equation (6.1) we see that at a higher temperature,
the last term, $-\Delta U/T$, becomes less important since the denominator becomes larger.[1] This negative contribution to ΔS_{tot}, which stems
from the reduction in spreading of energy in the surroundings, then
plays a smaller role compared to the first term, that mainly consists of
the positive entropy contribution from the increase in number of configurations. Since the negative entropy contribution $-\Delta U/T$, which
counteracts evaporation, becomes less important, the liquid evaporates more easily and the vapor pressure increases as T increases. In
Section 3.2, we expressed this fact in a different but equivalent way,

---

[1] If ΔU and ΔS are approximately constant independent of T, it is easy to realize
this. As we shall see later, this condition is, however, not necessary. The conclusion is
true – at least in a limited temperature range – even when ΔU and ΔS are temperature
dependent (see footnote 13).

namely, that the energy required for evaporation is more easily accessible from the surroundings at higher temperature. This follows from the fact that it becomes easier for the surroundings to deliver energy ΔU to the system since the entropy change of the surroundings, which equals $-\Delta U/T$ and is unfavorable, becomes smaller at increasing temperatures. In other words, the penalty to deliver energy is smaller at high temperatures (see also the optional material in Section 2.8).

As we raise the temperature further, the vapor pressure of the liquid continues to increase. An increasing proportion of the molecules will be present in the gas phase and since V is constant this gives rise to the increased pressure in the closed vessel. Let us assume that we initially have a sufficient amount of liquid in relation to the vessel's volume, so that all the liquid does not evaporate during the heating. We can, in fact, heat up the system far beyond the normal boiling point of the liquid and both gas and liquid will still be left in the vessel. The pressure in the vessel will then be far greater than normal atmospheric pressure.

The density of the vapor thus increases more and more as T is increased. The gas density eventually becomes so great that the interactions between the molecules in the gas phase become strong.[2] At the same time, the density of the liquid decreases slightly because it expands. The increase in density for the gas and the decrease for the liquid will continue with increasing temperature until we eventually reach a point where the gas and liquid densities are equal.[3] This point is called the **critical point**. For example, for water this occurs at $374°C$ and then the vapor pressure is 218 atm. At this point, there is no distinction between gas and liquid. At temperatures higher than the critical value, there is no separation of the system into a gas and a liquid phase (so-called phase separation). This applies irrespective of the gas pressure – a gas cannot condense into liquid when compressed

---

[2]Our simple reasoning about the configurational entropy et cetera then does not apply because the gas is far from ideal. Both the energy and the entropy of the gas phase depend on the intermolecular interactions. Equation (6.1) is, however, valid in the general case when V and T are constant.

[3]For this to occur, the amount of substance in relation to the vessel's volume must be exactly such that both phases are present all the time during heating. If there is too little liquid, it will evaporate before this point is reached. In case there is too little gas, it will be compressed so much due to the increasing pressure that it disappears and only liquid remains.

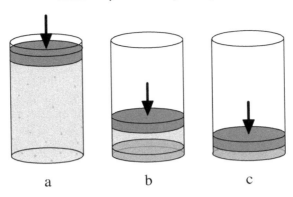

a   b   c

Figure 6.1   A cylinder containing a pure gas is compressed at a constant temperature T, which is below the critical value. Heat must be removed when compression is carried out in order to keep T constant. (a) When the gas pressure has become equal to the vapor pressure of the liquid at temperature T, the gas begins to condense into liquid. (b) When the volume is reduced further, more of the gas condenses into liquid. The pressure is constant and equal to the vapor pressure as long as both gas and liquid are present. (c) When all gas has condensed, only liquid remains. If the volume is reduced further, the liquid is compressed.

at temperatures above the critical value. The compression results only in a gas with increasingly high density.

However, if we compress a gas when the temperature is lower than the critical value, the gas will eventually condense into a liquid.[4] The condensation occurs when the gas pressure has increased to a value that is equal to the vapor pressure of the liquid at the temperature in question; recall that this is the pressure at which liquid and vapor are in equilibrium with each other. Initially, a few liquid droplets form on the vessel walls, as illustrated in Figure 6.1a. When we continue to compress the gas (by reducing the volume of the vessel) at constant temperature, the amount of fluid increases and the amount of gas decreases (Figure 6.1b). This is because the liquid has a higher density than the gas, so the system responds to the decrease in volume by gas turning into liquid. The gas pressure is constant and is equal to the vapor pressure as long as both gas and liquid are present in equilibrium with each other. When the volume has become so small that the

---

[4]Provided that the temperature is not so low that the gas instead solidifies.

entire vessel has just become filled with liquid (all gas has condensed to liquid, Figure 6.1c), the pressure will begin to increase again when the volume is further reduced. Since we are then compressing a liquid, the pressure will increase rapidly because the liquid is dense and its molecules repel each other when pressed together. One says that the liquid has a low compressibility; one has to expose it to high pressures in order to reduce its volume slightly.

After this interlude on gas compression, let us return to what happens when we change the temperature. We have considered the case of constant volume, but the case of constant pressure is more relevant in most applications. If we, for example, heat water on the stove, the pressure is constant (it is equal to the atmospheric pressure) and then something special happens when the temperature reaches 100°C, namely, that the water boils (provided we have normal atmospheric pressure in the kitchen). Boiling means that more and more liquid evaporates as we add heat, while the water temperature remains constant (at 100°C in this case) until eventually all the liquid has evaporated. This is true no matter how much liquid we have from the beginning. When no liquid remains, the temperature will rise above 100°C when we add more heat, but then the water is present only in the form of steam. Note that the process is different than when we kept the volume constant, in which case we could heat the liquid to much higher temperatures than 100°C provided sufficient liquid was present from the beginning.

Thus, there is an essential difference between the cases of constant volume and constant pressure. The fact that water boils at 100°C has to do with the circumstance that the pressure is equal to normal atmospheric pressure. If we would be located at a high mountain top where the pressure is lower, we would find that water boils at a lower temperature. How come? Why, in the first place, does a liquid boil when the temperature is high enough? To answer these questions, we will study the case of constant pressure in detail.

Let us place a small glass of liquid in a closed container filled with air. After a while, the air also contains liquid vapor. The container has a freely movable piston. On the outside of the piston the atmospheric pressure is constant, as illustrated in Figure 6.2, and therefore the total gas pressure P inside the container is kept constant. The system and the surroundings have temperature T, which we can vary. For each temperature, however, we hold T constant until equilibrium is

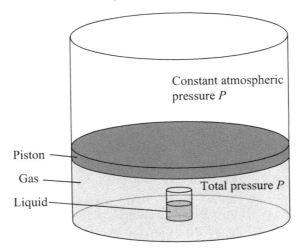

Figure 6.2   A glass with a liquid placed in a closed container which has a constant pressure P equal to the ambient pressure. The gas in the container consists of air and liquid vapor.

reached between the liquid and the vapor. We will now investigate evaporation at constant P and T.

Let $\Delta H = q$ be the heat (see Equation 4.29) that the system in the container takes up during the evaporation and ΔS be the change in entropy. The change in Gibbs energy is $\Delta G = \Delta H - T\Delta S$ and according to Equation (4.36) the total change in entropy of the system and the surroundings is

$$\Delta S_{tot} = -\frac{\Delta G}{T} = \Delta S - \frac{\Delta H}{T}, \tag{6.2}$$

where $-\Delta H/T$ is the entropy change of the surroundings. We have $\Delta S > 0$ and $\Delta H > 0$ for the process liquid $\rightarrow$ vapor. As we have seen in Sections 4.6 and 4.7, the fact that we have ΔG instead of ΔA and ΔH instead of ΔU (compare with Equation 6.1) means that we take into account that the volume of the system changes (the piston moves) to keep the total pressure P constant. When $\Delta G < 0$ ($\Delta S_{tot} > 0$) evaporation is spontaneous and when $\Delta G > 0$ ($\Delta S_{tot} < 0$) condensation is spontaneous, i.e., the reverse process (compare with Figure 4.15). When equilibrium is reached (that is, when G attains its minimum value at the current temperature), the partial pressure of vapor in the gas is equal to the vapor pressure of the liquid at the temperature in question.[5] In the same way as in the discussion of Equation (6.1) we can conclude that the vapor pressure increases with increasing temperature because the term $-\Delta H/T$ becomes less important.

Thus, the liquid vapor content of the gas inside the container grows with increasing temperature (we assume that sufficient liquid is initially in the glass so that there is always some liquid left). The gas volume becomes larger in order to maintain the total pressure P constant. Since the number of air molecules in the gas is unchanged, the partial pressure of air in the mixture decreases at the same time (an increasingly smaller percentage of the mixture consists of air).[6] The sum of the vapor pressure and the partial pressure of air inside the container remains equal to P. As long as the vapor pressure is less than the ambient atmospheric pressure P everything is fine. The piston is pushed out as much as needed to make the pressure on the inside and outside equal and pressure balance to occur.

Something new happens, however, when one heats up the system to the temperature at which the vapor pressure becomes equal to atmospheric pressure P. Then the sum of the vapor pressure and the partial pressure of air in the container must exceed P. The pressure inside the container thus becomes greater than the external pressure regardless of how far the piston is pushed out. Hence, pressure balance cannot be obtained and the piston will be completely pushed out from the container. (Even in the absence of air inside the container, this would be the case if T is increased just a tiny bit above this temperature – the vapor pressure then exceeds the atmospheric pressure P.) When the piston has been completely pushed out, any remaining liquid will be vaporized as heat is applied, i.e., the liquid boils. We have accordingly reached the boiling point. If the temperature becomes slightly higher than the boiling point during the boiling, the vapor pressure exceeds the ambient atmospheric pressure, whereby the steam bubbles that form will expand greatly and displace the surrounding atmosphere. When steam is formed, the required heat is taken from the liquid, which effectively prevents the temperature

---

[5]Actually, the partial pressure of vapor in the gas mixture at equilibrium, the so-called partial vapor pressure, is not exactly equal to the vapor pressure above a pure fluid at the same temperature (in the latter case there is no other substance than the vapor of the liquid present in the gas phase). The difference is usually negligible provided that the partial pressure of the gas present in addition to the vapor (i.e., air in our case) is not *very* high.

[6]We here ignore that a small portion of the air is dissolved in the liquid.

from increasing any further despite the fact that heat is added to the system all the time. This applies until all liquid has vaporized.

The **boiling point** of a liquid is accordingly the temperature at which its vapor pressure is equal to the atmospheric pressure. On top of a mountain where the atmospheric pressure is lower, the boiling point is therefore lower. In a pressure cooker the boiling point is instead higher, because the cooker has a valve with a spring mechanism that does not allow steam to pass out until the pressure inside is much greater than the external pressure. The higher boiling temperature makes the food in the cooker become cooked faster than at the normal boiling point.

Thus, the boiling point T_b depends on the pressure P, the fact of which we can highlight by writing $T_b(P)$. At the boiling point a pure liquid is in equilibrium with pure vapor (without air) at pressure P. This means that two pure phases (liquid and gas) at the same pressure are in equilibrium with each other. If the pressure is the same but the temperature is lower, the pure substance is in the form of liquid, and when the temperature is higher it is in the form of gas. Notice the difference between a pure substance (i.e., only one component) and a system where we have a mixture of gases present (for example, air and vapor). A pure liquid cannot be in equilibrium with pure vapor when the pressure is P and $T < T_b(P)$; the substance is solely in the form of liquid under these conditions. It is, however, possible for the liquid to be in equilibrium with a mixture of vapor and air when the *total* pressure is P and $T < T_b(P)$ in accordance with our discussion above. The partial pressure of vapor is then equal to the vapor pressure of the liquid (which is less than P); the remainder of the total pressure P is comprised of the partial pressure of air.[7]

Let us continue to discuss pure substances at constant pressure P, which condition we do not write out explicitly anymore. A transition from liquid to gas at the boiling point T_b is called a **phase transition**. Since the liquid and the gas are in equilibrium, Gibbs energy does not change during the transition, that is, we have $G^{gas} = G^{liquid}$. This fact is a consequence of the condition of equilibrium because if G were not unchanged during the transition, a spontaneous process would take place in the direction of reduced G, which would imply that the equilibrium does not prevail – contrary to what we have assumed.

---

[7]A small amount of air is dissolved in the liquid, which we can disregard in this context.

If we have one mole substance that changes from liquid to gas at the boiling point, we thus have $G_m^{gas} = G_m^{liquid}$ (index m means "per mole") and if n moles changes, we have $nG_m^{gas} = nG_m^{liquid}$. Generally, we can write the criterion for equilibrium between liquid and gas as follows:

$$G_m^{gas} = G_m^{liquid} \tag{6.3}$$

and we have

$$\Delta_{vap}G_m = G_m^{gas} - G_m^{liquid} = 0 \quad \text{when } T = T_b, \tag{6.4}$$

where $\Delta_{vap}G_m$, the Gibbs energy of vaporization, is defined by the first equality. Let us, in what follows, assume that we have one mole that changes from liquid to gas.

What about temperatures around T_b? How does G vary as a function of T, i.e., $G = G(T)$? We have $G = H - TS$ so if H and S were constants independent of temperature, G plotted as a function of T would be a straight line[8] with the slope $-S$. In reality, H and S are not constant, but if we restrict ourselves to a small temperature interval, $G(T)$ can still be approximated as a straight line with a slope of $-S$.[9]

Since the gas has more particle configurations than the liquid, we have $S^{gas} > S^{liquid}$. This means that G^{gas} plotted as a function of T has a larger (negative) slope than G^{liquid}, see Figure 6.3 where we have written subscript m to indicate that we consider each quantity per mole of substance. When $T < T_b$ we see that $G_m^{gas} > G_m^{liquid}$ so the system will reduce G when all the gas condenses into liquid, so the liquid is the stable state (equilibrium state). At higher temperatures, $T > T_b$, we have $G_m^{gas} < G_m^{liquid}$ and the gas is stable instead. When the temperature is gradually increased from below T_b to above, G for the system will assume the lowest possible value at each temperature all the time, i.e., G will follow the thick curve segments in Figure 6.3. Therefore, the substance is in liquid form below T_b, vaporizes at T_b, and is in gas form above T_b. If the temperature is reduced from a value above T_b, the substance instead condenses at T_b.

---

[8]A straight line has the equation $y = kx + l$ where k is the slope coefficient and l the intercept (intersection) with the y-axis, that is, $y = l$ when $x = 0$. In the present case we have $y = G$, $x = T$, $k = -S$, and $l = H$, provided S and H were constants independent of T.

[9]The fact that the slope is $-S$ even when H and S are temperature dependent is shown later.

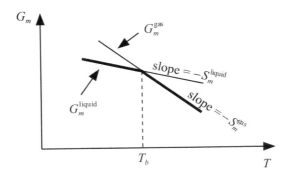

Figure 6.3   Gibbs energy per mole of pure liquid, G_m^{liquid}, and pure gas, G_m^{gas}, plotted as functions of temperature when the pressure is constant. The two curves intersect at the boiling point T_b and have slopes $-S_m^{\text{liquid}}$ and $-S_m^{\text{gas}}$, respectively. The thick curve segments show which G value is the lowest one at each temperature.

Let us follow what happens when we add heat to the system that contains solely a pure liquid at constant pressure. When $T < T_b$ the added heat is used to raise the temperature of the liquid. The heat required to raise the temperature by ΔT is given according to Equation (4.30) by $q = C_P^{\text{liquid}}\Delta T$, where C_P^{liquid} is the heat capacity C_P for the liquid. When we reach the boiling point, we need to add heat to vaporize the liquid. The amount of heat needed for this is according to Equation (4.29) equal to the difference in enthalpy between the initial state (liquid) and final state (gas), that is, H^{liquid} and H^{gas}, respectively. This difference,

$$\Delta_{\text{vap}}H = H^{\text{gas}} - H^{\text{liquid}} \tag{6.5}$$

is called the **enthalpy of vaporization** (or **heat of vaporization**) and is accordingly the heat we have to add to vaporize the liquid to gas (it is denoted $\Delta_{\text{vap}}H_m$ per mole of substance). During boiling, T remains constant until all liquid is vaporized. Thereafter, when $T > T_b$, the added heat is used to raise the temperature of the gas. An amount of heat equal to $q = C_P^{\text{gas}}\Delta T$ has to be supplied for the temperature to increase by ΔT, where C_P^{gas} is the heat capacity of the gas, which has a different value than that of the liquid (generally a lower value).

At the boiling point the entropy of the system will change from S^{liquid} to S^{gas} when we add heat. In Figure 6.3, we see this as a change in the slope at $T = T_b$ when we move along the thick curve segments. Since $\Delta G = \Delta H - T\Delta S$ and $\Delta_{\text{vap}} G = 0$ at $T = T_b$, we see that $\Delta_{\text{vap}} H - T_b \Delta_{\text{vap}} S = 0$, which we can write as

$$\Delta_{\text{vap}} S = \frac{\Delta_{\text{vap}} H}{T_b}, \tag{6.6}$$

where $\Delta_{\text{vap}} S = S^{\text{gas}} - S^{\text{liquid}}$ is the **entropy of vaporization**. This is exactly what the second law of thermodynamics says about entropy change for a reversible process at constant temperature, namely, $\Delta S = q/T$, where $q = \Delta H$ since P is constant. Provided that the evaporation takes place slowly by a gradual addition of heat, it is a reversible process at $T = T_b$ since liquid and vapor are at equilibrium for this temperature. At other temperatures evaporation (or condensation) is an irreversible process since it is spontaneous.

The corresponding applies when a solid melts to form a liquid, such as the phase transition from ice to water. The molecules in a solid (crystalline) phase sit attached at certain locations and in certain orientations in the crystal, while the molecules in a liquid are freely movable. Therefore, the entropy will increase when a solid melts, $S^{\text{liquid}} > S^{\text{solid}}$. In order to free the molecules from their attachments in the crystal, one has to add energy (heat) and therefore $H^{\text{liquid}} > H^{\text{solid}}$. When we plot G^{solid} and G^{liquid} as functions of T, the slope of the first graph is smaller than the second because $S^{\text{solid}} < S^{\text{liquid}}$, as shown in Figure 6.4 which is a sketch of how G for the various phases depends on temperature (we have written index m in the figure to indicate that the plotted G is per mole of substance). In reality, the graph of $G = G(T)$ for each phase is slightly curved, which is a consequence of the fact that H and S depend on T. The slope at each point (the derivative) equals $-S$ at the temperature in question.[10]

---

[10]When we plot $G = H - TS$ as a function of T when P and N are constant, $G = G(T)$, the slope is $dG/dT = -S$ even when the curve is not a straight line, that is, when we take into account that H and S are temperature dependent. This can be understood mathematically as follows: We have $dG/dT = dH/dT - d(TS)/dT$, where the derivative of the product TS is given by $d(TS)/dT = S + T\,dS/dT$. According to the second law of thermodynamics $dS = dq/T$, and when P is constant we have $dq = dH$, which means that $T\,dS = dH$ and hence $T\,dS/dT = dH/dT$. Therefore, $dG/dT = dH/dT - [S + T\,dS/dT] = -S$. (This implies that the temperature dependence of G is primarily due to the factor T in the term TS.) Incidentally, we note that the corresponding fact applies to the Helmholtz energy $A = U - TS$. The temperature derivative of A is $-S$ when V and N are constant, which can be shown in the same way since $dq = dU$ in this case.

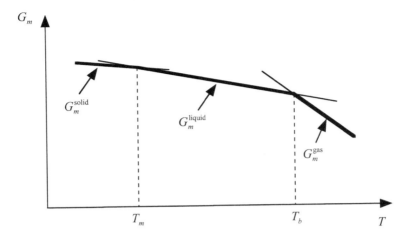

Figure 6.4 Sketch of Gibbs energy per mole for the solid, G_m^{solid}, liquid, G_m^{liquid}, and gas, G_m^{gas}, phases of a pure compound plotted as functions of temperature when the pressure is constant. The first two curves cross each other at the melting point T_m and the last two at the boiling point T_b. In reality, the graphs are slightly curved. The slope at any point is equal to $-S_m$ of the phase at the temperature in question. The thick curve segments show the G value that is lowest at each temperature. The region around T_b is also shown in Figure 6.3.

At the melting point T_m there is equilibrium between the solid and liquid phases, and accordingly $G^{\text{liquid}} = G^{\text{solid}}$. This means that the graphs for G of these phases in Figure 6.4 intersect at that temperature. When $T < T_m$ the solid phase has the lowest G and between T_m and T_b the liquid has the lowest G. This implies that the solid is the most stable phase below the melting point and the liquid is the most stable one when $T_m < T < T_b$. The heat needed to melt the solid at the melting point

$$\Delta_{\text{fus}} H = \left[H^{\text{liquid}} - H^{\text{solid}} \right]_{T=T_m} \tag{6.7}$$

is called the **enthalpy of fusion** (or **heat of fusion**). It is denoted $\Delta_{\text{fus}} H_m$ per mole of substance. The **entropy of fusion**, $\Delta_{\text{fus}} S$, is likewise the difference in entropy between the liquid and solid phases at the melting point. It satisfies

$$\Delta_{\text{fus}} S = \frac{\Delta_{\text{fus}} H}{T_m}, \tag{6.8}$$

since $\Delta_{fus}G = \Delta_{fus}H - T\Delta_{fus}S = 0$ when $T = T_m$ (compare with Equation (6.6)). At all other temperatures, $\Delta G = \Delta H - T\Delta S \neq 0$ for the process solid phase $\rightarrow$ liquid phase and the sign of ΔG determines which phase is stable in accordance with Figure 6.4. When $T < T_m$ the liquid freezes and when $T > T_m$ the solid melts spontaneously.

Accordingly, for both solid/liquid and liquid/gas at equilibrium, the Gibbs energy per mole, G_m, is equal in the two phases. As we saw when we obtained Equation (6.3), this expresses that G does not change when the substance goes from one form to the other, i.e., when the substance is transferred between the two phases in question. On the other hand, when G_m for the two phases are different, we found that the substance is transferred from the phase with the highest G_m to that with the lowest. Therefore, it is customary to introduce the concept of **chemical potential**, denoted μ, which can be used to determine to which phase a substance spontaneously "wants to go," that is, in which direction it shall be transferred for G to decrease. The chemical potential of a pure substance is simply equal to G_m, that is, $\mu = G_m$.[11] The substance is thus transferred from the phase of highest μ to that with the lowest, and at equilibrium, μ is equal in the two phases. The name "chemical potential" is used to show the similarity to other types of potentials, such as the gravitational potential (i.e., the potential energy in the gravitational field). A particle tends to move from a high to a low potential.

The condition of equilibrium between gas and liquid, Equation (6.3), can thus alternatively be written

$$\mu^{gas} = \mu^{liquid} \tag{6.9}$$

and that for equilibrium between liquid and solid phase

$$\mu^{liquid} = \mu^{solid}. \tag{6.10}$$

The curves in Figures 6.3 and 6.4 can, if desired, be denoted by μ instead of G_m and hence one plots $\mu = \mu(T)$ for the different phases.

---

[11] For a mixture, the chemical potential μ_i of a substance of species i is equal to the rate of change of Gibbs energy when the number of moles of that substance increases, while keeping the number of moles of the other species j in the mixture constant. (Mathematically, this is expressed as $\mu_i = (\partial G/\partial n_i)_{T,P,n_{j\neq i}}$.) Molecules of species i therefore go from a phase with high μ_i to one with low μ_i since G for the former phase then decreases more than G for the latter increases, so the total change of G is negative. At equilibrium μ_i for both phases is the same, which applies for all substances in the mixture.

Key points

- A pure gas that is compressed at a constant temperature (below the critical value) is condensed and turns into a liquid when the gas pressure becomes equal to the vapor pressure. The pressure is constant upon continued compression until all gas has condensed.

- A pure liquid boils at the temperature (the boiling point T_b) where its vapor pressure is equal to the ambient atmospheric pressure. When $T = T_b$ we have $G_m^{gas}(T) = G_m^{liquid}(T)$.

- A pure solid phase (for instance ice) melts at the temperature T (the melting point T_m) where $G_m^{solid}(T) = G_m^{liquid}(T)$.

- The chemical potential μ for a pure substance is equal to G_m. When two phases are present, the substance passes spontaneously from the phase of highest μ to that with the lowest. At equilibrium, μ is equal in the two phases.

6.2   It depends on the temperature
Temperature dependence of various quantities

6.2.1   *T* dependence of Gibbs energy

In the previous section we studied the phase transitions between solid and liquid (melting/freezing), and liquid and gas (evaporation/condensation). One of the questions we investigated was in which direction the transition occurs spontaneously at different temperatures when the pressure is constant. We found that we can determine this by studying $G = H - TS$, where we assumed, as an approximation, that H and S are constants independent of T. Likewise, we can study

$$\Delta G = \Delta H - T\Delta S \qquad (6.11)$$

for the transition, where we can assume that ΔH and ΔS are constants independent of T. The temperature dependence of ΔG is thus given mainly by the factor T in the $T\Delta S$ and if we plot ΔG as a function of T, we obtain approximately a straight line with slope $-\Delta S$. Alternatively, we can study

$$\frac{\Delta G}{T} = \frac{\Delta H}{T} - \Delta S \qquad (6.12)$$

(compare with Equation (6.2)). If we plot $\Delta G/T$ as a function of $1/T$ we obtain approximately a straight line[12] with slope ΔH.[13] Equations (6.11) and (6.12) both say that the entropy change of a system becomes more significant and the enthalpy change becomes less significant when T is increased. This means that high entropy is more important than low enthalpy at high temperatures, whereas the reverse is true at low temperatures. We took advantage of this fact in the last section in the discussion of the temperature dependence of vapor pressure and this is also the reason why one often observes the sequence solid $\rightarrow$ liquid $\rightarrow$ gas when the temperature rises. The solid phase has low enthalpy and low entropy so it is favored by low T, while gas has high enthalpy and high entropy so it is favored by high T.

The same principles apply to the temperature dependence of ΔG for other processes, for instance, chemical reactions. Equations (6.11) and (6.12) with constant ΔH and ΔS can generally be used as good approximations to determine how ΔG changes with T provided that we restrict ourselves to a sufficiently small temperature interval. We shall now use this fact for a particularly important case, namely, to determine how chemical equilibria depend on temperature.

6.2.2 T dependence of equilibrium constant

In Section 5.2, we found that $\Delta_r G^0$ (see Figure 5.6) plays a central role for chemical equilibria. The equilibrium constant K is given by Equation (5.15) which we can write, using Equation (6.12), as

$$\ln K = -\frac{\Delta_r G^0}{RT} = -\frac{\Delta_r H^0}{R}\frac{1}{T} + \frac{\Delta_r S^0}{R}, \tag{6.13}$$

where we have introduced $\Delta_r H^0$ and $\Delta_r S^0$, which are the changes in enthalpy and entropy, respectively, for the process depicted in Figure

---

[12]The straight line $y = kx + l$ has in this case $y = \Delta G/T, x = 1/T, k = \Delta H$, and $l = -\Delta S$, provided ΔH and ΔS are approximately constant independent of T.

[13]When we plot $G/T = H/T - S$ as a function of $1/T$ while P and N are constant, the slope at any point is equal to H even when the curve is not a straight line, that is, when we take into account that H and S are temperature dependent. This can be shown a similar way as the fact that the derivative of G with respect to T is equal to $-S$ when P and N are constant, which we showed in footnote 10. The same applies to $\Delta G/T$ whereby the slope is ΔH. Incidentally, we note that similar arguments apply for $\Delta A/T$ when V and N are constant, whereby the slope is ΔU (see the end of footnote 10). This result is used in footnote 1. Mathematically, we have $(\partial(\Delta G/T)/\partial(1/T))_{P,N} = \Delta H$, called the Gibbs-Helmholtz equation, and $(\partial(\Delta A/T)/\partial(1/T))_{V,N} = \Delta U$.

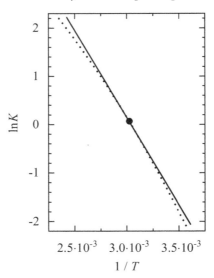

Figure 6.5   The dotted curve gives an example of the logarithm of the equilibrium constant as a function of $1/T$. It is approximated by the solid straight line that shows $\ln K$ according to Equation (6.13) with constant $\Delta_r H^0$ and $\Delta_r S^0$. The line is tangent to the curve at the solid circle which indicates the point where $\Delta_r H^0$ and $\Delta_r S^0$ were determined.

5.6. This equation gives the temperature dependence of the equilibrium constant, and it implies that if we plot $\ln K$ as a function of $1/T$, we obtain approximately a straight line with a slope of $-\Delta_r H^0/R$, as depicted in Figure 6.5. In the use of Equation (6.13) we can take the value of $\Delta_r H^0$ and $\Delta_r S^0$ at one temperature (marked by a filled circle in the figure), insert these values into the equation as constants and then calculate $\ln K$ at various temperatures T in the vicinity of this temperature. The straight line thus obtained is plotted in Figure 6.5.[14]

Equation (6.13) with constant $\Delta_r H^0$ and $\Delta_r S^0$ can be used to relate the equilibrium constant at two different temperatures T_1 and T_2. If we take the difference of $\ln K$ according to Equation (6.13) at the two

---

[14]The slope at any point on the curve is $-\Delta_r H^0(T)/R$ when $\Delta_r H^0$ depends on temperature. Therefore, this straight line is tangent to the curve at the temperature where $\Delta_r H^0$ was determined.

temperatures, we obtain after simplification

$$\ln K_2 - \ln K_1 = \frac{\Delta_r H^0}{R}\left[\frac{1}{T_1} - \frac{1}{T_2}\right], \tag{6.14}$$

where K_1 and K_2 are the equilibrium constants at temperatures T_1 and T_2, respectively. For an endothermic reaction, $\Delta_r H^0 > 0$ and if $T_2 > T_1$ the right-hand side of Equation (6.14) is positive, which means that $\ln K_2 > \ln K_1$ and hence $K_2 > K_1$. For endothermic reactions, the equilibrium constant accordingly increases with increasing temperature, which implies that the equilibrium will be shifted towards the products.[15] The products, which have higher enthalpy than the reactants because $\Delta_r H^0 > 0$, are thus favored at increased T. This result corresponds to what we found for the evaporation of a liquid, which is also an endothermic process: The vapor (the "product") is favored by an increased temperature compared to the liquid (the "reactant"). The reason is the same in both cases, namely, that the energy needed for the process becomes more easily available from the environment with increasing temperature. For exothermic reactions, the opposite is true; the equilibrium is shifted towards the reactants at higher temperatures. These consequences of shifts in temperature are further examples of Le Châtelier's principle, which we mentioned in Section 5.2.2.

6.2.3 T dependence of internal energy and enthalpy

When studying $\Delta_r G^0 = \Delta_r H^0 - T\Delta_r S^0$ or $\Delta_r G^0/T = \Delta_r H^0/T - \Delta_r S^0$ at different temperatures, it is, as we saw in Figure 6.5, a reasonable approximation to assume that $\Delta_r H^0$ and $\Delta_r S^0$ are constants independent of T. As we have utilized in Section 6.1, the same is true when studying $G = H - TS$, where one can assume as an approximation that H and S are independent of T. However, H and S *are* temperature dependent, so $H - TS$ is actually affected not only by the change in the factor T, but also by changes in H and S when T is varied. However, the changes in H and S give only a small contribution to G within a limited temperature interval; an increase in H is almost completely canceled by a corresponding increase in S in the combination

---

[15]This shift at increased K follows from Equation (5.32), where the products appear in the numerator and the reactants in the denominator; the amount of products thus increases and reactants decreases when K goes up.

$H - TS$.[16] Individually, however, H and S are changed most significantly when the temperature changes. First we look at the temperature dependencies of enthalpy and internal energy.

In Section 4.6, Equation (4.30), we saw that the change in enthalpy when the temperature is varied by ΔT is given by

$$q = \Delta H = C_P \Delta T \quad \text{when } P = \text{constant}, \tag{6.15}$$

where q is the added heat. The change in internal energy is according to Section 4.5, Equation (4.21), given by

$$q = \Delta U = C_V \Delta T \quad \text{when } V = \text{constant}. \tag{6.16}$$

If the temperature is increased by some other means than adding heat, for instance, by stirring with a propeller, the second equality in Equations (6.15) and (6.16), respectively, still applies: $\Delta H = C_P \Delta T$ when P is constant and $\Delta U = C_V \Delta T$ when V is constant (the number of particles of all species should be kept constant too in both cases).[17] All these results are valid provided C_P and C_V are constants independent of T, which is a good approximation if the temperature interval ΔT is not too large. If the heat capacities depend on temperature, Equations (6.15) and (6.16) apply only for a small temperature change dT and for the addition of a small quantity of heat dq. The general relationships therefore read

$$dq = dH = C_P dT \quad \text{when } P = \text{constant} \tag{6.17}$$

and

$$dq = dU = C_V dT \quad \text{when } V = \text{constant}. \tag{6.18}$$

---

[16]This is a consequence of the following: When a small amount of heat dq is added, H and S are changed by dH and dS which satisfy $dS = dq/T = dH/T$, where we used the second law of thermodynamics and the fact that $dH = dq$ when the pressure is constant. Thus we have $dH = TdS$. If the temperature increases by dT when heat is added, $H - TS$ will change to $(H + dH) - (T + dT)(S + dS)$, which we can write $H - TS + dH - TdS - SdT - dTdS$. The last term is the product of two small numbers and can be neglected compared to the other terms. Thus $H - TS$ has been changed approximately by the amount of $dH - TdS - SdT$, which is equal to $-SdT$ since $dH = TdS$. In other words, the change of $H - TS$ originates mainly from the increase in the factor T, while the contributions dH and $-TdS$ cancel each other.

[17]For an ideal gas these two latter relationships apply, as we have seen in Section 4.6, irrespective of whether P or V are constant or not. This is a special case, so the conditions of constant P and V, respectively, are important in the general case.

However, in this book we assume throughout that the heat capacity is independent of T to a sufficient approximation, so Equations (6.15) and (6.16) can be used to calculate ΔH and ΔU when temperature is changed.[18]

6.2.4 T dependence of entropy

Finally, let us examine how the entropy depends on temperature T. We first assume that the volume V and the number of molecules N are constant. We increase T by increasing the energy U of the system. Since V and N are constant, we do this by adding an amount of heat q (the work w is zero). According to Equation (6.16), the temperature change becomes $\Delta T = q/C_V$. When we add energy, the number of ways to distribute the energy between the molecules will increase rapidly, as we have seen in Section 2.6. Thus, the number of available microstates, Ω, increases for the system. Each molecule has a large number of quantum states and we have to include all possible ways of distributing the energy, that is, all possible combinations of quantum states of the molecules with the given total energy, exactly as discussed in Section 2.6. Since Ω increases when U increases, the entropy $S = k_B \ln \Omega$ increases too, but the question is how much?

We first examine an ideal monatomic gas where the energy is solely translational. The number of microstates Ω of the system depends on U, V, and N, which is denoted $\Omega = \Omega(U, V, N)$. Let us recapitulate the V dependence, which we already know. For ideal gases, we found in Section 2.3 that Ω increases when we increase the volume. According to Equation (2.2) we have $\Omega = \mathcal{K} V^N$ where $\mathcal{K}$ is independent of V. For a macroscopic system, N is a very large number, so

---

[18]For a large change in temperature, $\Delta T = T_{\text{after}} - T_{\text{before}}$, one has to consider that heat capacity depends on temperature. Mathematically, by integrating Equations (6.17) and (6.18) one obtains the general expressions

$$q = \Delta H = \int_{T_{\text{before}}}^{T_{\text{after}}} C_P(T)dT \quad \text{when } P = \text{constant}$$

and

$$q = \Delta U = \int_{T_{\text{before}}}^{T_{\text{after}}} C_V(T)dT \quad \text{when } V = \text{constant},$$

where it is explicitly shown that $C_P(T)$ and $C_V(T)$ depend on temperature. When C_P and C_V are constant in the temperature interval they can be taken out from the integrand and Equations (6.15) and (6.16) are obtained.

Ω increases very rapidly with V. When we change the energy U for a monatomic gas, a similar result applies (this is shown in Appendix E)[19]

$$\Omega = \mathcal{K}' U^{3N/2}, \tag{6.19}$$

where $\mathcal{K}'$ is independent of U.[20]

Let us now change the energy from U_{before} to U_{after} while holding V and N constant. From Equation (6.19) we see that

$$\frac{\Omega_{after}}{\Omega_{before}} = \frac{\mathcal{K}' U_{after}^{3N/2}}{\mathcal{K}' U_{before}^{3N/2}} = \left[\frac{U_{after}}{U_{before}} \right]^{3N/2}.$$

Since $S = k_B \ln \Omega$ this implies (compare with the derivation of Equation (2.8) in Section 2.4)

$$
\begin{aligned}
\Delta S &= k_B \ln \Omega_{after} - k_B \ln \Omega_{before} = k_B \ln \frac{\Omega_{after}}{\Omega_{before}} \\
&= k_B \ln \left(\left[\frac{U_{after}}{U_{before}} \right]^{3N/2} \right).
\end{aligned}
$$

The change in entropy when the energy varies from U_{before} to U_{after} is therefore

$$\Delta S = \frac{3}{2} N k_B \ln \frac{U_{after}}{U_{before}} \quad \text{(monatomic ideal gas).} \tag{6.20}$$

In Section 2.7, we saw that the temperature T is related to how the entropy changes with energy. Equation (2.14) says that $1/T$ (which is dS/dU) gives the rate of the entropy increase when we increase the energy (assuming that V and N are constant). Since Equation (6.20) is a relationship between entropy and energy, one can use this equation to determine the temperature of a monatomic gas that has a given energy. As we will show in the derivation below, Equation (6.20) implies that the temperature and the energy are proportional to each other;

---

[19]This expression is valid provided U is well above the ground state energy, which is taken as the zero level for U.

[20]If we combine these two results, we find that $\Omega(U,V,N) = \mathcal{K}'' U^{3N/2} V^N = \mathcal{K}''[U^{3/2}V]^N$ where $\mathcal{K}''$ is independent of U and V ($\mathcal{K}''$ depends only on N and on the molecular mass of the molecules). A complete expression for $\ln \Omega$ is given in footnote 9 of Appendix E.

more specifically, $U = \frac{3}{2}Nk_BT$. This is identical to Equation (4.19) of Section 4.4 that we obtained from a completely different reasoning there.

A LITTLE DERIVATION*

If U is changed by a small increment dU from $U_{\text{before}} = U$ to $U_{\text{after}} = U + dU$, Equation (6.20) gives the corresponding change dS in entropy

$$dS = \frac{3}{2}Nk_B\ln\frac{U+dU}{U} = \frac{3}{2}Nk_B\ln\left(1 + \frac{dU}{U}\right).$$

Since $\ln(1 + x) \approx x$ when x is a small number (see Figure 4.12), this implies that[21]

$$dS = \frac{3}{2}Nk_B\frac{dU}{U}. \tag{6.21}$$

Equation (2.14) implies that $dS = dU/T$ when V and N are constant (compare to Equation (2.18) with $dq = dU$ and $dS_{\text{irrev}} = 0$). If we insert this into the left-hand side of Equation (6.21), we obtain

$$\frac{dU}{T} = \frac{3}{2}Nk_B\frac{dU}{U}.$$

By dividing both sides by dU and rearranging the expression, we obtain

$$U = \frac{3}{2}Nk_BT \quad \text{(monatomic ideal gas),} \tag{6.22}$$

which is identical to Equation (4.19). Note that this relationship applies only to monatomic ideal gases since Equation (6.20) is valid for that case only.

As we noted in Section 4.5 in connection with Equation (4.22), the relationship $U = \frac{3}{2}Nk_BT$ means that we can identify the heat capacity $C_V = \frac{3}{2}Nk_B$ and we can therefore write Equation (6.20) as $\Delta S = C_V\ln(U_{\text{after}}/U_{\text{before}})$. Since U and T are proportional to each

---

[21]This can alternatively be obtained by differentiating $S = S(U) = \frac{3}{2}Nk_B\ln U +$ constant, which is valid provided U is well above the ground state energy.

other, we have $U_{\text{after}}/U_{\text{before}} = T_{\text{after}}/T_{\text{before}}$ and we obtain

$$\Delta S = C_V \ln \frac{T_{\text{after}}}{T_{\text{before}}} \quad \text{when } V = \text{constant}, \tag{6.23}$$

which describes how the entropy changes with temperature. Equation (6.23) is a very important result that we so far have shown only for monatomic ideal gases, but as we shall now see, this equation applies generally when C_V is constant independent of temperature.

First, let us increase the temperature of a system by a small amount dT from T to $T + dT$ by adding a small amount of heat dq when V and N are constant. We require that we always have equilibrium, so the heating takes place reversibly. According to Equation (6.18) we have $dq = C_V dT$. Equation (2.18) with $dS_{\text{irrev}} = 0$ says that when we add heat dq, the entropy increases by a small amount dS given by $dS = dq/T$. Thus, we have

$$dS = \frac{dq}{T} = \frac{C_V}{T} dT. \tag{6.24}$$

We seek the total entropy increase ΔS when the temperature is changed from T_{before} to T_{after}. When we repeatedly increase T incrementally by the amount dT, the entropy increase is given by the sum of the contributions dS for each step according to Equation (6.24).[22] The denominator T will become larger and larger, so the increase dS for each step therefore becomes smaller and smaller as the temperature increases, as illustrated in Figure 6.6. The temperature increase dT in each step should in principle be infinitely small ($dT \to 0$) and the number of steps infinitely many. Then, the total entropy increase ΔS between T_{before} and T_{after} is given by area under the curve C_V/T in the figure, that is,

$$\Delta S = \int_{T_{\text{before}}}^{T_{\text{after}}} \frac{C_V}{T} dT = C_V \int_{T_{\text{before}}}^{T_{\text{after}}} \frac{1}{T} dT, \tag{6.25}$$

where we have taken C_V outside the integration sign since it is assumed to be independent of T. To perform the integration we note that the integral is of the kind $\int_a^b (1/x)dx$ with $x = T$, $a = T_{\text{before}}$ and

---

[22] Anyone who is familiar with integration can go directly to Equation (6.25) by integrating Equation (6.24) between T_{before} and T_{after} to obtain the total entropy increase ΔS shown in Equation (6.23).

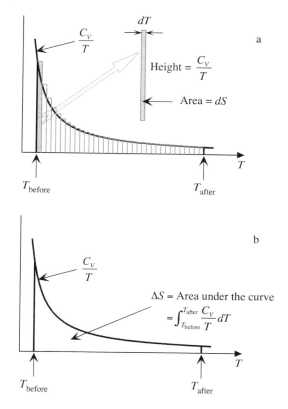

Figure 6.6 (a) The entropy increase, when T increases gradually with the increment dT in many small steps from T_{before} to T_{after}, is given by the sum of the contributions $dS = (C_V/T)dT$ for temperatures between these two values. These contributions are illustrated with a succession of narrow rectangles with width dT and height C_V/T. The first rectangle is gray toned and is also shown separately. When we reduce the width dT of each rectangle ($dT \rightarrow 0$) and simultaneously increase the number of rectangles between T_{before} to T_{after}, the sum of the surfaces of the rectangles will approach the area under the curve C_V/T. (b) The area between the curve and the T axis from T_{before} to T_{after} equals the total entropy increase ΔS.

$b = T_{after}$. Since

$$\int_a^b \frac{1}{x} dx = [\ln x]_a^b = \ln b - \ln a = \ln \frac{b}{a}$$

we obtain the same result as in Equation (6.23).

Equation (6.23) is thus a general result for the entropy change when the temperature changes and the volume is kept constant (assuming C_V is constant in the temperature interval).

One may now ask: How large is the entropy change when the pressure is kept constant instead of the volume and temperature changes? Let us first examine the case when we add a small amount of heat dq. According to the second law of thermodynamics $dS = dq/T$ in this case too (see Equation (2.18) with $dS_{irrev} = 0$) and according to Equation (6.17) we have $dq = C_P dT$ when the pressure is constant, which means that Equation (6.24) is replaced by

$$dS = \frac{dq}{T} = \frac{C_P}{T} dT. \tag{6.26}$$

Before we proceed, we shall, however, examine why this simple relationship is appropriate.

DISCUSSION OF EQUATION (6.26)

As we saw in Sections 4.5 and 4.6, the volume of a gas increases when we add heat dq at constant pressure and hence a part of the added energy is used to expand the gas against the ambient pressure (by pressing out the piston in the examples we discussed there). Therefore, a smaller amount of the added energy is available to be distributed among different quantum states of the molecules, which implies that the corresponding entropy increase (dS_{ener}) becomes smaller compared to the case of constant volume. At the same time, the number of possible particle configurations of the molecules increases as V increases, resulting in a positive entropy contribution (dS_{conf}). The total entropy change dS is the sum of these two contributions.

The expansion of the gas is reversible (the pressure of the gas is at all times equal to the external constant pressure) and we saw in Section 4.3 that the entropy change due to reversible work is zero. The loss of entropy of the system because of a part of the added energy is lost to the surroundings during the expansion, is therefore equal to the gain in configurational entropy (compare with the derivation of the ideal gas law in Section 4.4 where this fact was used). The total entropy change is therefore given by the total energy supplied in the form of heat dq according $dS = dq/T$, which is in accordance with what we wrote in Equation (6.26). The energy loss in the form of the reversible expansion work $dw = -PdV$ has no effect on the entropy of the system. On the other hand, this loss implies that the temperature increase in the system will be less than if the same amount of heat would be added to the system at constant volume, just as we found in Section 4.5. The temperature change is according to Equation (6.17) given by $dT = dq/C_P$ and we saw in Section 4.5 that $C_P > C_V$, which expresses the smaller increase in temperature.

Thereby, we have clarified the background of Equation (6.26) and we will use it to calculate the entropy change ΔS when the temperature is varied from T_{before} to T_{after} at constant P. Just like in the earlier case where we used Equation (6.24) to obtain Equation (6.23), we obtain

$$\Delta S = C_P \ln \frac{T_{after}}{T_{before}} \quad \text{when } P = \text{constant}, \tag{6.27}$$

where we have assumed that C_P is independent of temperature. The difference between the two cases "constant pressure" and "constant volume" is accordingly only which heat capacity one uses, C_P or C_V.

Key points

- The temperature dependence of G at constant P and N is given by $G = H - TS$, where one can assume, as an approximation within a small temperature range, that H and S are constant. When G is plotted as a function of T one thus obtains approximately a straight line with slope $-S$ in a limited temperature interval.[23]

- The temperature dependence of the equilibrium constant K is given at constant pressure by

$$\ln K = -\frac{\Delta_r H^0}{R} \frac{1}{T} + \frac{\Delta_r S^0}{R},$$

---

[23]When one considers that S and H depend on temperature, $S = S(T)$ and $H = H(T)$, the slope for $G = G(T)$ is still equal to $-S$ at each temperature. This is shown in footnote 10. The plot of $G = G(T)$ gives in general a curved line with second derivative $-C_P/T$ (this follows from $d^2G/dT^2 = -dS/dT = -C_P/T$, where the last equality is a consequence of Equation 6.26). In a sufficiently small T interval the curvature of the line can be ignored as an approximation. Mathematically, $(\partial G/\partial T)_{P,N} = -S$ and $\left(\partial^2 G/\partial T^2\right)_{P,N} = -C_P/T$.

where one can assume as an approximation within a limited temperature interval that $\Delta_r H^0$ and $\Delta_r S^0$ are constant independent of T. If one plots $\ln K$ as a function of $1/T$ one obtains approximately a straight line with slope $-\Delta_r H^0/R$.

- When the volume is constant and the temperature is changed from T_{before} to T_{after}, the internal energy and the entropy are changed by

$$\Delta U = C_V \Delta T \quad \text{and} \quad \Delta S = C_V \ln \frac{T_{after}}{T_{before}},$$

respectively, where $\Delta T = T_{before} - T_{after}$ and C_V is assumed to be independent of T.

- At constant pressure, the changes in enthalpy and entropy are

$$\Delta H = C_P \Delta T \quad \text{and} \quad \Delta S = C_P \ln \frac{T_{after}}{T_{before}},$$

respectively, where C_P is assumed to be independent of T.

CHAPTER 7

Epilogue

7.1 What are the molecules doing?

In this book, we have seen that by using our imagination and "taking part" in the world of the molecules, we can gain an understanding of various phenomena that we can observe in our world, i.e., on the macroscopic level. A large variety of properties and processes of macroscopic systems can be understood on the basis of what is happening on a molecular, microscopic level. The microscopic world is teeming with activity: molecules dart around back and forth, vibrate and rotate, they collide, exchange energy, repel and attract each other, and change shape, break apart and merge. All this is done continuously and everywhere. The individual molecules have different energies at each moment, they move, rotate, and vibrate at different rates, and have, for example, different shapes – all these properties change all the time for each molecule.

It is actually quite remarkable that this complex multitude of activities can give rise to macroscopic properties and processes that can be described in a relatively simple manner. The reason why this is possible is the fact that a macroscopic system consists of an extremely large number of molecules. On the macroscopic level, we usually experience averages of properties that involve a huge number of molecules and on a time scale that is relatively long from the molecular perspective. The fates of individual molecules play a very small role compared to what happens to the collective at large.

When a system has reached equilibrium, no changes take place as seen from a macroscopic perspective. The molecular concentrations do not change and, for example, the temperature, pressure, and total energy of the system remain constant. From the perspective of the individual molecules, on the other hand, a lot of things happen – the same things that happened before equilibrium was reached. The molecules dart around, collide, exchange energy, and are transformed. The difference between the conditions at equilibrium and what happens before equilibrium is reached (i.e., during a spontaneous process) has to do with the probabilities that various courses of events occur. At equilibrium, the possible courses are equally likely as their opposites. For example, on average, an equal number of molecules of each species is formed and disappears (that is, is converted into something else), the same amount of energy enters each part of the system as it leaves and as many molecules move in one direction as in the other. Before equilibrium is reached, it is, however, more likely that some courses of events take place than their opposites. The spontaneous process that one then perceives macroscopically is the net change that occurs in the direction that is most probable.

In thermodynamics one mainly studies equilibrium states and processes between the initial and the final equilibrium states. The initial state is thereby the equilibrium state that exists before the process is allowed to start; for example, before one lets heat pass between a hot and a cold body, before a barrier is removed that separates different molecular species from each other, before a candle is lit in the presence of air, or before the volume is allowed to change for a system that is exposed to an external force. The final state one considers may be a real or an imaginary equilibrium state; the latter is the case when one is interested in determining whether an imagined process can occur spontaneously. One does not necessarily need to be concerned with what happens in detail during the process – by comparing the initial and final states, one can still determine whether the process is possible. If the total entropy increases, the process is thermodynamically possible; if it decreases, the process is impossible.

The fact that a process is thermodynamically possible does not necessarily mean that it can happen spontaneously within a reasonable time. For some chemical reaction to occur, one may need to have a suitable catalyst present (a catalyst is a substance that increases the speed of a reaction without being consumed). On the other hand, if the process is thermodynamically impossible, it cannot happen spontaneously no matter which catalyst one adds. The reverse process is, however, possible and is speeded up by the catalyst.

The reason why some thermodynamically possible processes are slow is usually that molecules or their constituent atoms must pass a high energy barrier between the initial and final states. The barrier can be passed only for molecules that have high enough energy and at each moment of time there are just a few molecules that fulfill this. A catalyst makes the barrier become lower, whereby more molecules have sufficient energy to pass and the process occurs more rapidly. The initial and final states are, however, the same in both cases.

The probability we are talking about in connection with the distinction between spontaneous and nonspontaneous processes concern the probability that a process will take place during a time period that is sufficiently long. What is sufficient depends on the system – in some cases it can be a very long time, in other cases short. For instance, the fact that diamond is converted to graphite at room temperature and normal atmospheric pressure means that this is much more likely than the reverse taking place (graphite is the thermodynamically stable form). Yet, the probability of conversion of diamond to graphite is extremely small and the process takes a tremendous amount of time (in practice, it does not occur). The probability of a spontaneous process to take place is thus always much greater than the reverse process, irrespective of whether the first one is large or small.

For each final state that has been obtained in reality, the total entropy of the system and its surroundings has increased (in this connection we disregard reversible changes that must be performed infinitely slowly). A spontaneous process occurs, of course, in the direction that is most probable. This corresponds, as we have seen, to an increase in entropy. The link between probability and entropy is given by: (a) entropy equals $S = k_B T \ln \Omega$ where Ω is the number of microstates for a system with a given energy and (b) all microstates with the same energy are equally probable.[1] The equilibrium state is the macroscopic state that corresponds to the largest number of microstates and thereby the highest entropy and the highest probability. All other possible macroscopic states, for instance, the state that the system had before the spontaneous process was allowed to happen, correspond to much fewer microstates and hence lower entropy and lower probability.

Since a very large number of molecules is involved, the equilibrium state corresponds to the largest number of microstates; this number is enormously greater than for other possible macroscopic

---

[1] That all accessible microstates with the same energy are equally likely at equilibrium is one of the basic postulates of statistical mechanics. If the system is isolated, the energy is constant. All possible microstates of the system thus have the same energy and are equally likely. If the system can exchange energy, the total energy (for the system *and* the surroundings) is constant and the possible microstates of the system and its surroundings (considered as *one* system with constant energy) have the same energy and are thus equally likely.

states. Thereby, the system and the surroundings get access to a vastly
larger number of microstates when the system goes from the initial to
the final state. Although it is possible in principle for the process to
go in the reverse direction, this is extremely unlikely – in practice this
does not happen. For example, it is possible that all air molecules in a
room at one moment in time would spontaneously gather in one half
of the room, but the probability for this is so tremendously small that
this does not happen in practice. Even very small deviations from a
uniform distribution throughout the room are in fact very unlikely.
Equalization of the density is a spontaneous process because it is far
more likely that this happens than the reverse.

All processes that we observe in reality are spontaneous.[2] For example, a warm system (A) in cold surroundings will spontaneously
deliver heat to the surroundings if heat is allowed to be exchanged.
Thereby, the total entropy increases until the temperature of A and
the surroundings is the same. To heat up system A again, we must,
for example, bring A in contact with a body (B) with a higher temperature. The heat transfer *to* system A is thereby spontaneous and
the total entropy increases. To carry out this process, we must, however, first heat up body B by adding energy to it from some source.
During this preparation, which is also a spontaneous process, the total entropy increases. In each step that takes place, the total entropy
increases because energy is spread out more and more.

If we want to perform a process that cannot occur spontaneously,
we have to drive it forward by using another process that can occur
spontaneously. For example, a mixture of oxygen and hydrogen gases
(oxyhydrogen gas) reacts spontaneously to form water upon ignition

---

[2]There are developments in nature which are not spontaneous processes in the
sense that we use in this book. The motion of the moon around the earth is an example of one, and also the movement of a body which is not affected by any forces
and the velocity of which therefore does not change. These are virtually unchange-
able movements; the moon continues to circle around the earth and the body will
move straight forward for a very long time. Since the moon and the earth mutually
deform each other – partly in the form of tides on earth – the speed of the moon is,
however, slowly decreasing. *This change* is a spontaneous process in the sense that we
use. For similar reasons, a ball that bounces vertically up and down against the floor
in vacuum, will eventually stop due to friction, whereby the potential and kinetic
energy of the ball is converted mainly into thermal motion of the molecules. This is
a spontaneous process; the final state is the ball at rest on the floor. If the ball would
bounce perfectly elastically (without friction), it would continue to do so indefinitely.
This elastic bouncing is not a spontaneous process in our sense, and it corresponds to
the moon circling the earth.

or in contact with a catalyst. The driving force is the dispersion of the released energy. To do the reverse process – to split water into oxygen and hydrogen gases – we must add energy. For example, we can do this by electrolysis, whereby energy is supplied by an electrical current. This current must be generated by some process that is spontaneous. We could, for instance, use a generator driven by hydropower (water falls spontaneously from high to low altitude and thereby drives the generator) or by an engine that burns oil. The entropy increase when generating the electrical power is thereby greater than the reduction in entropy when producing oxygen and hydrogen gases from water. Overall, the total entropy therefore increases.

Often, we have a situation where the surroundings of the system participate only by making one or more variables of the system remain constant, such as the temperature. As we have seen, in order to keep track of the total entropy (of the system and the surroundings) in an efficient way, one introduces the concept of free energy. Free energy is so designed that its change is proportional to the change in total entropy, for instance, $\Delta A = -T\Delta S_{tot}$ at constant T and V. The free energy, in this case the Helmholtz energy $A = U - TS$, contains only the system's own variables: U = its energy, T = its temperature (which is also the temperature of the surroundings), and S = its entropy.

While the condition $\Delta S_{tot} > 0$ is a general criterion for spontaneous processes, the corresponding condition $\Delta A < 0$ is valid *only* for spontaneous processes at constant T and V. In the relationship $\Delta S_{tot} = -\Delta A/T = -\Delta U/T + \Delta S$, the term $-\Delta U/T$ represents the entropy change in the surroundings. This entropy change arises because of the heat transfer that takes place in order to keep the temperature of the system unchanged. For a spontaneous process, the free energy decreases and equilibrium is reached when it has become as small as possible. The equilibrium condition is that A has a minimum $(dA = 0)$ when T and V are constant. At the same time, S_{tot} assumes its largest possible value under the given conditions (that is, at constant T and V).

If the pressure P is kept constant, whereby V is allowed to vary, it is instead the change in Gibbs energy $(G = H - TS = U + PV - TS)$ that is proportional to the change in total entropy: $\Delta G = -T\Delta S_{tot}$ at constant T and P. Under these conditions, $\Delta G < 0$ is the criterion for spontaneous processes $(\Delta S_{tot} > 0$ obviously also applies). In this case, we have $\Delta S_{tot} = -\Delta G/T = -\Delta H/T + \Delta S$, where $-\Delta H/T$ is the term that expresses the entropy change in the surroundings due to heat exchange. The equilibrium condition at constant T and P is that G has its minimal value ($dG = 0$). S_{tot} assumes at the same time its largest possible value under the given conditions (that is, at constant T and P).

The enthalpy, $H = U + PV$, is so designed that it takes into account the energy exchange with the surroundings that takes place because the volume changes in order to keep P constant. In the relationship $\Delta H = \Delta U + P\Delta V$, which applies when P is constant, the term $P\Delta V$ constitutes the energy that the system delivers in the form of work on the surroundings when the volume changes. Heat added to the system, which is equal to ΔH when P is constant, goes partly to a change in energy of the system's molecules, ΔU, and partly to the work $P\Delta V$ carried out on the surroundings. These effects are thus included in both the enthalpy H and Gibbs energy G when P is constant.

Thermodynamics thus provides the means to determine, for example, which way a process can go spontaneously under various conditions, and what are the corresponding conditions for equilibrium. The fact that an extremely large number of molecules is involved in processes for macroscopic systems, makes macroscopic thermodynamics applicable without necessarily taking into account any details about what happens to the molecules. This is the strength of classical thermodynamics; it applies to macroscopic systems regardless of the molecular description. Since the microscopic world lies behind what is happening on the macroscopic level, the molecular description gives, however, an increased insight into and understanding of thermodynamics.

Thermodynamics was, however, developed during the 19th century, which was a time when molecular properties were largely unknown – several prominent scientists even doubted their existence. The reasons why thermodynamics was developed at that time was to a large extent due to an interest in energy conversions, such as conversion of heat to work in steam engines. That a general theory was developed under these conditions – a theory that at a later stage could also be applied to understand and describe molecular properties and behaviors – remains a truly great intellectual achievement.

Today, large areas of science are molecularly oriented and the molecular world is becoming better and better understood. The purpose of this book has been to make this world even more accessible by explaining what is the driving force in the world of molecules and a variety of other issues. So ... Welcome back to the world of molecules again. I hope that you feel at home there!

APPENDIX A

Heat dispersion and temperature, an analogy

A.1 Spreading of energy in a body

Molecules that interact are constantly exchanging energy with each other. At each instant in time the molecules have different energies – which ones have high and which ones have low energies is due to chance. A molecule with high energy can at the next moment have low and vice versa.

Figure A.1   People in a room who move around and occasionally meet each other. Every person has an amount of coins. When they meet, each one gives a random number of coins to the person she meets and accepts the money she receives. The persons represent molecules and the number of coins that each one has corresponds to the molecule's energy. The picture shows only a small part of the room. *Illustration: Anette Hedberg.*

Let us use an analogy: A large number of people are moving around in a room. They have been given a large amount of coins (Figure A.1), which all have the same value. The people represent molecules and the number of coins that each person has corresponds to the energy of a molecule. We assume that the people are immensely

Figure A.2   One of the persons has been given a gift of one million coins. This corresponds to a large amount of energy has been added to one molecule. *Illustration: Anette Hedberg.*

generous and share their coins with each other when they meet. Sometimes they give away one, sometimes two, and sometimes a larger number of coins completely at random. At the same time they receive coins from those they encounter.

At every point in time, different persons have a different number of coins, but on average, measured over a period of time, each and every one will have an equal number (averaged over a sufficiently long time). At a single moment, some have more and some less coins than the average, and a few may have a very large number of coins. It is even possible – but not very likely – that a single person at one instant in time has all the coins.

Let us now give a single person in the room one million coins, that is, much more than each one of the others. She and everyone else will, just as before, hand out and receive money in a random manner (Figure A.2). In the beginning, she will give away more than she receives. Those who she meets have significantly less money and cannot give away more than what they currently have. She can, however, give large amounts. Eventually, she will on average have the same amount of money as everyone else. The gift that she received has become dispersed among all the people.

Everyone has become richer on average. It is very unlikely (but not impossible) that she at a later time by chance will have her original one million coins back. A fairly even distribution is vastly more probable than an uneven distribution, where one or a few persons have far more or far less money than the rest.

At the molecular level, the reasoning above corresponds to what happens if we heat a specific part of a body by adding energy (heat). The energy spreads spontaneously: the heated molecules gradually lose most of the added energy while the energy of all other molecules is increased on average. The end result is that the entire body has become somewhat warmer than it was initially. The added energy has a very high probability of being distributed uniformly over all molecules.

A.2 Spreading between a hot and a cold body

Let us now examine the transfer of heat between a hot and a cold body – first when both bodies consist of the same molecular substance, and then when the bodies are different. Also in this case we make an analogy with humans.

Suppose we have two rooms with people who behave in exactly the same way as in our previous example. The two rooms correspond to the two bodies, the people represent the molecules that the bodies consist of and the value of money corresponds to the energy. The rooms are equally large and contain an equal number of persons. Initially the rooms have no connection with each other so the coins that are exchanged between the persons in each room stay there. Suppose also that the persons in one of the rooms are on average richer than in the other one. Room A, which contains the rich persons, corresponds to the hot body and room B to the cold body.

What happens when we open up gaps in the wall between the rooms so that money can be exchanged between rooms A and B? The persons themselves cannot pass through the openings. We assume that the persons in both rooms are just as willing to give and receive money through openings as they are when they meet inside the rooms. Coins flow in both directions because people, when they meet through the gaps, give coins to one another in a random manner (Figure A.3). More coins flow from room A to room B than vice versa, since the persons in A initially had more. Thus there is a net flow from A to B, whereby the people in A are becoming less wealthy on average, and in B less poor.

Also in this case, an equal distribution of coins among everyone becomes the most likely outcome. This is vastly more probable than

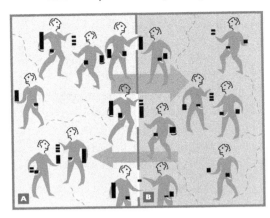

Figure A.3   People who are all equally generous are located in two rooms, A and B. The rooms represent two bodies composed of the same substance. Initially, the persons in A have on average more money than those in B. When gaps have been opened in the wall between the rooms, the giving and receiving of money also takes place between the rooms. Initially, the flow of coins is greater from room A to room B than the reverse (arrows). Room A represents a body which initially is hotter than the other one. Both rooms are larger than what is shown in the picture. *Illustration: Anette Hedberg.*

an uneven distribution where a number of persons have a lot more or a lot less money than the others. In other words, the flow of coins continues until all have become equally rich on average. Once this occurs, the flow of coins from A to B is on average equal to that from B to A. The net flow is zero.

We can express this by using a kind of "money temperature," which initially must be higher in A because A spontaneously delivers and B receives a net flow of coins when the gaps are opened. When the net flow is down to zero, the temperature is the same in the two rooms.

This example corresponds to the transfer of heat from a hot to a cold body. The number of ways to distribute energy about evenly between the molecules is much larger than the number of ways to distribute the energy unevenly. The reason for the heat transfer is accordingly that it is much more likely with a uniform distribution than a nonuniform.

A.3   Lower temperature – but higher energy

Let us now consider two bodies consisting of different substances. We will again make an analogy with people in two rooms.

What happens if the people in room A and B do not behave in the same way? Say that the B persons are inclined to keep some of the money for themselves. Let us assume that they have several pockets where they put money. Each B person takes up a number of coins from one pocket, puts some of it in another of her own pockets and gives the rest to the person she meets. An A person, on the other hand, always gives away everything she happens to pick up from her pocket. The A and B persons represent different kinds of molecules with unequal abilities to contain energy; they correspond to two molecular species with different numbers of quantum states per energy level.

Let us further assume that the A persons initially, just as before, are richer on average than their neighbors in B. When we open gaps in the wall, coins begin to flow in both directions (Figure A.4), and in the beginning the flow from room A to room B is, of course, greater than from B to A.

The net flow does, however, not become zero when the A and B persons after a while have equally many coins each on average. The

Figure A.4   Here, the people in room B are a little stingy, while those in A are exactly like before. The B persons have several pockets where they put money. The rooms represent two bodies consisting of different molecular substances. Initially, the persons in A have on average more money than those in B and the flow is greater from room A to B than the reverse. *Illustration: Anette Hedberg.*

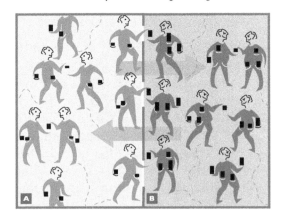

Figure A.5   When the flow from B to A finally is equal to that from A to B, equilibrium is reached and the "money temperature" is equal in the two rooms. Then, each one of the B persons, who are a bit stingy, has more money on average than each A person. This illustrates how molecules of two different substances typically have different energy even though the temperature is equal. *Illustration: Anette Hedberg.*

"cheapskates" in room B continue, as usual, to put coins in their own pockets. Therefore the net flow from A to B continues until the B persons have significantly more coins per person than their neighbors in A – so many more that each B person, despite her stinginess, on average gives away as much as each A person. Only at that stage does the net flow between the rooms become zero (Figure A.5).

Also in this case, the "money temperature" in room A is initially higher than in B, since A is spontaneously delivering and B receiving a net flow. Moreover, the temperature continues to be higher in A when the point has been reached where the A and B persons are equally rich on average. The spread of coins continues to increase when even more coins are transferred from A to B. The number of ways to distribute the coins becomes greater because the persons in room B distribute the money not only between each other but also between their various pockets.

Eventually, the persons in room B have become significantly richer, the net flow has become zero and the temperature has accordingly become equal. Just before this equilibrium has been reached, the temperature of B is still lower than of A, despite that the number of coins per person is higher in B.

A body may thus have a lower temperature than another body even though the energy per molecule is greater in the former. It should be emphasized that this is normal for bodies that are composed of different kinds of molecules.

The condition that the temperature is higher in one body (A) than in another (B) means that it is more likely that energy (heat) flows from hot to cold than in the opposite direction when contact occurs. The flow is due to a spreading of energy. A net amount of energy is transferred between the bodies as long as the total number of ways to distribute the energy can increase. The number of possible ways to distribute the energy in A decreases (energy is removed), while the number in B increases (energy is supplied).

While the total number of possible distributions (for A and B together) increases during the energy transfer, the transfer occurs spontaneously. No further net transfer takes place when the total number of ways to distribute the energy has reached its maximum and cannot increase any more upon a continued transfer. Thereby, the most probable state is reached, and the temperature is the same in the two bodies.

Accordingly, the temperature of a body has no simple relationship to the energy per molecule.[1] It is instead associated with how the number of possible distributions of energy for a body (the number of microstates) changes when its energy is varied. The exact relationship between temperature and the change in the number of distributions is explained in the following box, which provides a mathematical supplement to the discussion of temperature in Section 2.7 (the contents of the box are, however, not needed in order to read the rest of the book).

TEMPERATURE AND THE RELATIONSHIP TO ENTROPY*

Let system A have $\Omega_A(U_A)$ and system B have $\Omega_B(U_B)$ microstates when their energies are U_A and U_B, respectively. The total number of microstates for systems A and B together is according to Equation (2.4) equal to $\Omega_{AB} = \Omega_A(U_A)\Omega_B(U_B)$. Let a small amount of energy (heat) dU pass from A to B, so the energy of A becomes $U_A - dU$ and of B $U_B + dU$ (we assume that $dU > 0$). If Ω_{AB} then increases, the

---

[1] However, for certain parts of the energy, such as the translational energy, there are simple relationships (see Section 4.4).

heat transfer is spontaneous and A must by definition have a higher temperature than B. When Ω_{AB} increases, $\ln \Omega_{AB}$ increases too and therefore also

$$S_{AB} = k_B \ln \Omega_{AB} = k_B \ln \Omega_A \Omega_B = k_B \ln \Omega_A + k_B \ln \Omega_B = S_A + S_B.$$

Now

$$dS_A = \frac{dS_A(U_A)}{dU_A} dU_A = \frac{dS_A(U_A)}{dU_A}(-dU)$$

and

$$dS_B = \frac{dS_B(U_B)}{dU_B} dU_B = \frac{dS_B(U_B)}{dU_B} dU$$

so the total increase in entropy is

$$dS_{AB} = dS_A + dS_B = \left[-\frac{dS_A(U_A)}{dU_A} + \frac{dS_B(U_B)}{dU_B} \right] dU > 0. \qquad (A.1)$$

Since $dU > 0$ the bracket must be positive and therefore

$$\frac{dS_A(U_A)}{dU_A} < \frac{dS_B(U_B)}{dU_B}. \qquad (A.2)$$

The result in Equations (A.1) and (A.2) means that the system, which receives heat (B) and thereby increases its entropy, must have a larger increase in S due to the energy exchange than the reduction in S of the other system (A). This is natural, because otherwise the total entropy $S_A + S_B$ and therefore Ω_{AB} would not increase, contrary to our starting point. According to Equation (A.2), the system with the *lowest* temperature has the *largest* value of the derivative dS/dU. By taking

$$T = \frac{1}{dS/dU}$$

as a measure of temperature, a warm body will have a higher temperature than a cold one. This is the definition of absolute temperature introduced in Section 2.7.

The Boltzmann distribution law*

In this appendix we proceed with the arguments in Section 2.8 one step further in order to determine the probability that system A at a constant temperature T (Figure 2.27) is in a particular microstate with energy U_A. We denote this probability $p_A(U_A)$. The temperature is determined by the surroundings B. Just as in Section 2.8 it is assumed that B is very much greater than A, so that the temperature is not affected by the energy exchange between A and B.

Let the total energy of the combined system AB be equal to $U_{AB} = U_A + U_B$, where U_{AB} is constant while U_A and U_B may vary because of energy transfer between A and B. The entire system AB is assumed to be isolated and does not exchange any energy with its environment. At equilibrium, each of the microstates of AB is equally probable in accordance with what has been said earlier (Section 2.6) and we denote this probability p_{AB}.[1] When A is in a particular microstate with energy U_A, the number of possible microstates for the entire AB is equal to $1 \times \Omega_B(U_{AB} - U_A)$ because B can be in any of $\Omega_B(U_{AB} - U_A)$ different states (compare with the discussion of the first row of Figure 2.15 in Section 2.4).

The probability for each one of the microstates of AB is equal to p_{AB} regardless of the distribution of energy between A and B, so p_{AB} is a constant independent of U_A. Among all microstates of AB there are $\Omega_B(U_{AB} - U_A)$ which have system A in a particular microstate with energy U_A. Therefore the probability of observing this state is $\Omega_B(U_{AB} - U_A)$ times greater than p_{AB}, that is, $p_A(U_A) = \Omega_B(U_{AB} - U_A)p_{AB}$. Thus, $p_A(U_A)$ is proportional to $\Omega_B(U_{AB} - U_A)$. According to Equation (2.24) with $U = U_{AB}$ and $\Delta U = U_A$ we have

$$\Omega_B(U_{AB} - U_A) = \Omega_B(U_{AB})e^{-\frac{U_A}{k_B T}},$$

where $\Omega_B(U_{AB})$ is constant since U_{AB} is constant. Hence, $\Omega_B(U_{AB} - U_A)$ is proportional to $\exp(-U_A/k_B T)$ and it follows that the same thing is

---

[1] We have $p_{AB} = 1/\Omega_{AB}$ because the total number of microstates of AB is Ω_{AB} and all of them are equally probable.

true for $\mathfrak{p}_A(U_A)$, so we have

$$\mathfrak{p}_A(U_A) = \mathcal{K}_A \times e^{-\frac{U_A}{k_B T}}, \tag{B.1}$$

where $\mathcal{K}_A$ is a constant of proportionality. This constant can be determined in the following manner. If we add up a probability over all possibilities, we obtain the result 1 and therefore

$$1 = \sum_{\text{all microstates of A}} \mathfrak{p}_A(U_A) = \mathcal{K}_A \times \sum_{\text{all microstates of A}} e^{-\frac{U_A}{k_B T}} = \mathcal{K}_A \times Z_A,$$

where we have summed over all microstates of A (for all possible energies U_A) and where Z_A denotes the sum

$$Z_A = \sum_{\text{all microstates of A}} e^{-\frac{U_A}{k_B T}}. \tag{B.2}$$

Thus, $\mathcal{K}_A = 1/Z_A$ and we have[2]

$$\mathfrak{p}_A(U_A) = \frac{e^{-\frac{U_A}{k_B T}}}{Z_A}. \tag{B.3}$$

This relationship is called **Boltzmann's distribution law** and it is of great importance in statistical mechanics. As we have seen in Section 2.8, the exponential function in the numerator is a measure of the availability of energy from the environment at temperature T. The part of $\mathfrak{p}_A(U_A)$ that is specific to system A is included in the denominator Z_A, which contains a sum over all possible microstates of A.[3]

---

[2]From the reasoning above and footnote 1 we can see that $\mathcal{K}_A = \Omega_B(U_{AB})\mathfrak{p}_{AB} = \Omega_B(U_{AB})/\Omega_{AB}$, but the advantage with Equation (B.2) is that it provides an expression for $\mathcal{K}_A = 1/Z_A$ in terms of the properties of system A alone.

[3]Incidentally, it can be mentioned that in statistical mechanics one makes the identification $A_A = -k_B T \ln Z_A$, with Z_A from Equation (B.2) – a very important relationship. Here, A_A is Helmholtz energy for system A, which is introduced in Section 3.8. A key property of Helmholtz energy is that when some process occurs in A at constant temperature T and volume V we have $\Delta A_A = -T \Delta S_{\text{tot}}$ (see Equation 3.7), where S_{tot} is the total entropy of system A and its surroundings (here system B is the surroundings, so $S_{\text{tot}} = S_{AB}$). It is easy to see that $A_A = -k_B T \ln Z_A$ indeed has this relationship to S_{tot}. From footnote 2 we see that $\ln \mathcal{K}_A = \ln \Omega_B(U_{AB}) - \ln \Omega_{AB}$ which implies that $k_B \ln \mathcal{K}_A = S_B(U_{AB}) - S_{AB}$. Therefore, $-k_B T \Delta \ln Z_A = k_B T \Delta \ln \mathcal{K}_A = -T \Delta S_{AB} = -T \Delta S_{\text{tot}}$ since $S_B(U_{AB})$ is constant and vanishes in the difference (it is the entropy of B in the absence of A).

APPLICATION: DISTRIBUTION OF MOLECULAR SPEEDS*

One application of the Boltzmann distribution law is to let system A be composed of a single particle in an ideal gas, and let B be the rest of the gas, which we assume is a macroscopic system at a given temperature T. The gas particles can exchange energy with each other and they have different speeds at different moments of time. We shall determine the probability that a particle has the speed v. The kinetic energy (translational energy) for a molecule with speed v is $\varepsilon_{tr} = mv^2/2$. We are here only considering that the particle has translational energy.[4]

Let us begin with motions in the x direction. The velocity component in this direction is v_x and the corresponding kinetic energy is $mv_x^2/2$. According to the Boltzmann distribution law, the probability of observing a microstate with energy ε is proportional to $\exp(-\varepsilon/k_BT)$. It is not meaningful to talk about the probability that a particle has exactly the velocity v_x, but instead we ask ourselves what is the probability that the velocity is between v_x and $v_x + dv_x$, where dv_x is a small number. For the motion in the x direction, the energy is $\varepsilon = mv_x^2/2$ and the probability that the particle has velocity between v_x and $v_x + dv_x$ is thus proportional to $\exp(-mv_x^2/2k_BT)dv_x$. The reason why we have the factor dv_x here is that the probability must be proportional to dv_x (if we increase dv_x with, say, a factor of 2, the probability must be twice as large).

We are now ready to treat motions in three dimensions: in the x, y, and z directions. We start with the probability that the velocity is between v_x and $v_x + dv_x$ in the x direction at the same time as it is between v_y and $v_y + dv_y$ in the y direction and between v_z and $v_z + dv_z$ in the z direction (see Figure B.1a). This probability is proportional to

$$e^{-\frac{mv_x^2}{2k_BT}} dv_x \times e^{-\frac{mv_y^2}{2k_BT}} dv_y \times e^{-\frac{mv_z^2}{2k_BT}} dv_z$$

---

[4]If the particle also has any other energy such as rotational energy or internal binding energy, one can show that this does not affect the probability that the particle has a certain speed. When the energy is $\varepsilon_{tr} + \varepsilon_{other}$ where ε_{other} is any other form of energy (which is independent of the particle's velocity), then $\exp(-[\varepsilon_{tr} + \varepsilon_{other}]/k_BT) = \exp(-\varepsilon_{tr}/k_BT) \times \exp(-\varepsilon_{other}/k_BT)$. This means that the probabilities for the distribution of translational energy and for the other energy are independent of each other (the proportionality constant in the Boltzmann distribution can also be written as a product for the same reason).

since the three probabilities in each direction are independent of each other. The probability that the velocity of a particle is equal to $\mathbf{v} = (v_x, v_y, v_z)$ within the margin (dv_x, dv_y, dv_z) is thus equal to

$$\text{const.} \times e^{-\frac{m(v_x^2 + v_y^2 + v_z^2)}{2k_BT}} dv_x dv_y dv_z,$$

where "const." is a proportionality constant and where we have used the exponentiation rule $e^a e^b e^c = e^{a+b+c}$. This probability is accordingly proportional to the exponential function times the volume $dv_x dv_y dv_z$ of the small box in Figure B.1a and the expression gives the probability for a particle motion in a certain direction. In the figure, the "tip" of the velocity vector then lies within the volume $dv_x dv_y dv_z$ of the box.

However, we are not interested in what direction the particle has, but only in its speed v independently of the direction of $\mathbf{v}$ (see Figure B.1b). We want to find the probability that the particle's speed is between v and $v + dv$. In Figure B.1b, the tip of the velocity vector is therefore allowed to be within the volume of a spherical shell of radius v and thickness dv. Since dv is a very small number, which we assume here, the shell has a volume equal to the shell's area times its thickness, i.e., $4\pi v^2 \times dv$.

The probability $P(v)dv$ that the speed of a particle lies between v and $v + dv$ regardless of direction is thus obtained by replacing $dv_x dv_y dv_z$ with $4\pi v^2 dv$ in the expression above. We thereby obtain

$$P(v)dv = \text{const.} \times e^{-\frac{mv^2}{2k_BT}} 4\pi v^2 dv, \qquad (B.4)$$

where we have used that $v^2 = v_x^2 + v_y^2 + v_z^2$ according to the Pythagorean theorem. The proportionality constant can be determined from the fact that the sum of the probability of all possibilities must be equal to 1, meaning that the integral of $P(v)$ over all speeds v is equal to 1. One can show that it means that the constant must be $(m/2\pi k_BT)^{3/2}$ so we have

$$P(v) = \left(\frac{m}{2\pi k_BT}\right)^{\frac{3}{2}} e^{-\frac{mv^2}{2k_BT}} 4\pi v^2. \qquad (B.5)$$

This result is called the **Maxwell-Boltzmann distribution law** for

molecular speeds and it is plotted in Figure 2.1 for different temperatures T. Since the temperature and the particle mass occur only as
m/T in this expression, one can realize that if one increases m by, say,
a factor of 2, one obtains the same distribution as if one instead reduces the temperature by the same factor. Figure 2.1 therefore also
shows the distribution of speed for particles with different masses at
the same temperature, where the curve for low temperature then corresponds to high mass and high temperature corresponds to low mass.

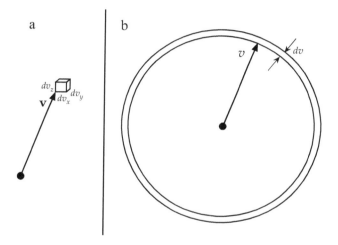

Figure B.1 (a) When the velocity is between v_x and v_x+dv_x in the x direction
at the same time as it is between v_y and v_y+dv_y in the y direction and between
v_z and v_z+dv_z in the z direction, the "tip" of the velocity vector $\mathbf{v} = (v_x, v_y, v_z)$
lies within the depicted box with sides dv_x, dv_y and dv_z. (b) When the speed
$v = |\mathbf{v}|$ lies between v and $v+dv$ regardless of direction, the tip of the velocity
vector $\mathbf{v}$ lies within the depicted spherical shell of thickness dv.

Collision with a piston in motion*

In this appendix we shall consider what happens when a particle collides with the surface of, for example, a piston. We take for granted that the piston has a much larger mass than the particle and we assume for simplicity that the surface is completely smooth. Let us begin with the case of a stationary piston. Consider a particle that approaches the surface. It is appropriate to divide the particle velocity vector $\mathbf{v}$ into a component perpendicular to the surface, v_x, and two components along with the surface, v_y and v_z (see Figure C.1). The speed $v = |\mathbf{v}|$ is given according to the Pythagorean theorem by $v^2 = v_x^2 + v_y^2 + v_z^2$. At the collision, the particle's motion will change direction. If the collision is elastic, the speed will be the same as before the collision. The only thing that happens is that v_x changes sign (see Figure C.2).

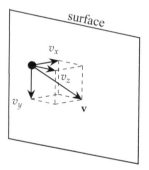

Figure C.1   A particle that approaches a surface. The velocity vector $\mathbf{v}$ of the particle has been divided into components, v_x, v_y, and v_z. We have chosen the x axis perpendicular to the surface and the y and z axes along the surface.

Let us now assume that the piston moves with speed u to the left towards the approaching particle (see Figure C.3a). If we were to consider the particle from the surface's point of view, that is, if we would accompany the piston in its motion, the particle would approach with the speed $v_x + u$ in the x direction (see Figure C.3b).

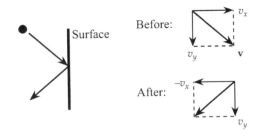

Figure C.2   In an elastic collision between a particle and the surface of a stationary body that is very heavy, the sign of the particle's velocity component perpendicular to the surface is changed while the other velocity components are unchanged (the z component is perpendicular to the plane of the paper and is not shown). In all figures, v_x denotes the initial value of the x component of the velocity.

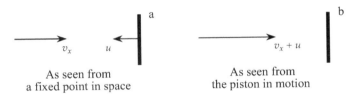

Figure C.3   (a) A particle approaches a piston that is moving to the left with speed u. When viewed from a fixed point in space, both the particle and piston are moving. (b) From the piston's perspective, the particle approaches with the relative speed $v_x + u$ in the x direction while the piston itself has the speed zero.

At the collision, the particle will change the sign of the velocity component in the x direction when we consider the collision as seen from the surface, which thus is stationary relative to ourselves (see Figure C.4a). As seen from a fixed point in space, however, the particle has not only changed the sign of the velocity's x component, but the magnitude of this component differs by $2u$ from the original (see Figure C.4b). Thus, speed of the particle has increased from $v_{\text{before}} = [v_x^2 + v_y^2 + v_z^2]^{1/2}$ to $v_{\text{after}} = [(v_x + 2u)^2 + v_y^2 + v_z^2]^{1/2}$. This is what happens when one compresses a gas, whereby the speeds of the particles increase when they collide with the piston.

Let us now turn to the case where the piston moves with speed u to the right, i.e., in the same direction as the approaching particle (see

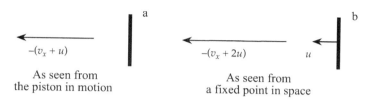

As seen from
the piston in motion

As seen from
a fixed point in space

Figure C.4   (a) After the collision, the particle has changed the sign of its velocity in the x direction when viewed from the piston's perspective. (b) As seen from a fixed point in space, the particle velocity in the x direction differs by $-u$ from the relative velocity in (a), that is, its magnitude is $2u$ larger than it was before the collision.

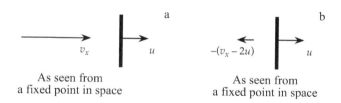

As seen from
a fixed point in space

As seen from
a fixed point in space

Figure C.5   (a) A particle approaches a piston that moves to the right with speed u. (b) After the collision, the magnitude of the particle velocity in the x direction has decreased by $2u$ compared to what it was before the collision. (If v_x initially is less than $2u$ but greater than u, the particle does not change direction at the collision, but the magnitude of the velocity component is changed in the same way.)

Figure C.5a; only particles moving towards the piston with at least speed u in the x direction will collide with the surface). The treatment is completely analogous to the previous case, but one must change u to $-u$ all formulas (see Figure C.5b), and we leave the analysis as an exercise to the reader. The particle speed decreases from v_{before} to $v_{after} = [(v_x - 2u)^2 + v_y^2 + v_z^2]^{1/2}$. This is what happens when one expands a gas, whereby the speeds of the particles are reduced when they collide with the piston.

Kinetic energy and pressure*

Consider a gas enclosed in a container. Let us examine the gas molecules' collisions with the container walls. The force acting on the wall at a collision depends on how fast the molecule approaches the surface. A fast molecule causes a larger force than a slow molecule. It is the velocity component perpendicular to the surface, v_x, which is crucial (we choose the x axis perpendicular to the surface). This component changes sign at the collision (see Figure C.2). The force on the wall also depends on the mass of the molecule; a twice as heavy molecule gives rise to twice the force. The relevant quantity is the product mv_x, which is called momentum. According to the laws of mechanics, the force is equally large as the change in momentum per unit time. At the collision, the momentum changes from mv_x to $-mv_x$ in the x direction, i.e., a decrease by the amount $2mv_x$. Thus the contribution from each collision is proportional to mv_x.

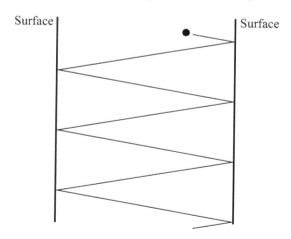

Figure D.1 A molecule in an ideal gas collides repeatedly with the surfaces of the container walls.

The total force on each surface depends on the number of collisions that takes place on the surface per unit time. A fast molecule will collide more often than a slow molecule; each molecule will, of course, sooner or later return to the wall as we can realize from Figure D.1.

The number of times per second that the molecule collides with a surface is proportional to the velocity component v_x. Therefore, the contribution to the force *from each molecule* will be proportional to the square of v_x:

$$\text{Force} = \text{const.} \times \underbrace{m \times v_x}_{\text{From each collision}} \times \underbrace{v_x}_{\text{Several collisions}} \qquad \text{(D.1)}$$

Since different molecules have different speeds, the total force on the wall is proportional to the mean value of the force from the individual molecules

$$\text{Total force} = \mathcal{B} \times m \left\langle (v_x)^2 \right\rangle, \qquad \text{(D.2)}$$

where $\langle \cdot \rangle$ denotes the average over all molecules and $\mathcal{B}$ is a new constant. The total force on the surface is, of course, proportional to the surface area a and to the molecular density $\rho = N/V$ of the gas, so these factors are included in $\mathcal{B}$. A more detailed analysis shows that $\mathcal{B}$ contains solely these two factors (readers who are particularly interested are referred to the derivation at the end of this appendix), so we have $\mathcal{B} = a\rho$ and hence

$$P = \frac{\text{Total force}}{a} = \rho m \left\langle (v_x)^2 \right\rangle. \qquad \text{(D.3)}$$

The speed v for each individual molecule is according to the Pythagorean theorem given by

$$v^2 = (v_x)^2 + (v_y)^2 + (v_z)^2,$$

where v_x, v_y, and v_z are velocity components in the x, y, and z directions, respectively. If we take the average over all molecules, we obtain

$$\left\langle v^2 \right\rangle = \left\langle (v_x)^2 \right\rangle + \left\langle (v_y)^2 \right\rangle + \left\langle (v_z)^2 \right\rangle.$$

For a macroscopic gas phase there is no difference between properties of the gas in the x, y, and z directions, so therefore the average value of the squared velocity component must be equal in all directions

$$\left\langle (v_x)^2 \right\rangle = \left\langle (v_y)^2 \right\rangle = \left\langle (v_z)^2 \right\rangle. \qquad \text{(D.4)}$$

This implies that

$$\langle v^2 \rangle = 3\langle (v_x)^2 \rangle \tag{D.5}$$

and therefore we have according to Equation (D.3)

$$P = \frac{\rho m \langle v^2 \rangle}{3}. \tag{D.6}$$

Each molecule has a kinetic energy (translational energy) equal to $\varepsilon_{tr} = mv^2/2$, so the mean value of the translational energy per molecule is $\bar{\varepsilon}_{tr} = \langle \varepsilon_{tr} \rangle = m\langle v^2 \rangle/2$.[1] Thus we have the pressure due to the molecular collisions

$$P = \frac{2\rho\bar{\varepsilon}_{tr}}{3}, \tag{D.7}$$

which is our final results (used in Section 4.4, Equation (4.17)).

DERIVATION OF THE CONSTANT B IN EQUATION (D.2)*

At each collision with the surface, a molecule's velocity in the x direction changes from v_x to $-v_x$, that is, a change with an absolute value of $2v_x$. The corresponding change in momentum is $2mv_x$. If l is the distance between the two surfaces in Figure D.1, the molecule will collide with the right surface $v_x/2l$ times per second (the distance that the molecule traverses in the x direction between two such collisions is $2l$). During one second, the total change in momentum in the x direction due to the molecule's collisions with right surface is hence $2mv_x \times v_x/2l = mv_x^2/l$ in absolute value.

The change in momentum per second is according to the laws of mechanics equally large as the force. Thus, each molecule affects the surface with the force mv_x^2/l, which means that the proportionality constant in Equation (D.1) is $1/l$. The volume between the surfaces is la and the number of molecules there is equal to $la\rho$. The total force on the surface from all molecules is the number of molecules $la\rho$ times the average of the force mv_x^2/l per molecule, that is, the force is $a\rho\langle mv_x^2 \rangle$. In Equation (D.2), we accordingly have $B = a\rho$.

---

[1]The notations $\bar{\varepsilon}_{tr}$ and $\langle \varepsilon_{tr} \rangle$ mean the same thing. The former is, however, more compact to write and is therefore used in the main text.

APPENDIX E

Kinetic energy and entropy for monatomic gas*

In this appendix, we will examine an ideal monatomic gas, the energy of which only consists of kinetic energy (translational energy). The gas is enclosed in a box of volume V. A molecule's energy is given by $\varepsilon_{tr} = mv^2/2$, where m is the mass of the particle, $v = |\mathbf{v}|$ its speed, $\mathbf{v} = (v_x, v_y, v_z)$, and v_x, v_y, and v_z are the velocity components in the x, y, and z directions, respectively. We have $v^2 = (v_x)^2 + (v_y)^2 + (v_z)^2$ according to the Pythagorean theorem. It is appropriate here to introduce the concept of momentum that is defined as $\mathbf{p} = m\mathbf{v}$, so the momentum components are given by $p_\alpha = mv_\alpha$ for $\alpha = x, y, z$. The translational energy is

$$\varepsilon_{tr} = \frac{(p_x)^2 + (p_y)^2 + (p_z)^2}{2m} \tag{E.1}$$

expressed in these components.

The energy for each molecule is quantized and can only assume discrete values. This quantum mechanical phenomenon can (somewhat improperly) be expressed as that only certain values of momentum components are possible for a monatomic molecule that bounces between the walls.[1] A quantum-mechanical treatment, which we will not do here,[2] shows that there is a minimal absolute value, p_{min}, for

---

[1] In quantum mechanics the momentum p_α for $\alpha = x, y, z$ is quantized for a particle in a box. A particle with a well-defined momentum p_α does not have a velocity in the classical sense, but it is described by a so-called standing wave between the walls of wavelength $\lambda = h/|p_\alpha|$, where h is Planck's constant. The standing wave may be said to describe the particle which bounces between the walls. Similarly to the vibrations of a guitar string that is stretched between two walls, the standing wave can only have certain wavelengths and therefore only certain values of p_α are possible.

[2] For those who are familiar with the quantum mechanical treatment of what is often called "the particle in a box," the following may serve as a reminder (see also footnote 1). We have a gas of noninteracting particles in a box that we, for simplicity, assume to be cubic with sides L, so the box volume is $V = L^3$. The sides of the box are aligned with the coordinate axes. (That we choose a cubic box makes no difference for the macroscopic properties of a homogeneous gas since these properties do not

Figure E.1   The possible absolute values of p_x, p_y, and p_z for a monatomic molecule in a box are represented by vertical bars (which continue infinitely upwards). Bold lines show each component's value for the molecule in the example.

each component (where $p_{min} \neq 0$) and that all other possible values are integer multiples of p_{min}. For instance, $|p_x|$ can assume the values p_{min}, $2p_{min}$, $3p_{min}$, et cetera, and the same thing applies for p_y and p_z. Figure E.1 shows an example where[3]

$$|p_x| = 4p_{min}, |p_y| = 5p_{min} \text{ and } |p_z| = 3p_{min}.$$

Another possibility is, for instance, that

$$|p_x| = 5p_{min}, |p_y| = 4p_{min} \text{ and } |p_z| = 3p_{min},$$

where $4p_{min}$ and $5p_{min}$ have changed places. The number of possibilities is, as we can realize, very large. Each constitutes a possible quantum state of the monatomic molecule.[4]

For a gas with N molecules, there are a total of $3N$ momentum components (three per molecule), which generally have different values.[5] An example with three molecules is given in Figure E.2. Each

---

depend on the shape of the box that the gas is contained in.) Consider, for example, the momentum p_x in the x direction for one of the particles. The quantization of p_x is due to the fact that the standing wave (the solution to the Schrödinger equation) must have zero amplitude at the surfaces of the box walls. This means that the available wavelengths $\lambda = h/|p_x|$ satisfy the condition $j\lambda/2 = L$, where j is any positive integer (like for a guitar string), and we thus obtain $|p_x| = jh/2L$. The smallest possible positive integer is 1, so the minimum value of $|p_x|$ is $p_{min} = h/2L = h/(2V^{1/3})$. This means that $|p_x| = jp_{min}$ for any integer $j \geq 1$. The same applies in the y and z directions.

[3]As explained in footnote 2, p_{min} for the x, y, and z directions can, without loss of generality, be assumed to be the same for a macroscopic, homogeneous gas.

[4]The internal quantum states of the molecule (electronic states) are not considered here. We only include those associated with the translational energy.

[5]If the energy is sufficiently high, it is very unlikely that two components have the same value because the number of possible values is much larger than the number of molecules.

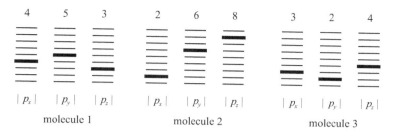

Figure E.2   Illustration of three molecules with examples of values of the momentum components. The numbers at the top show the value of each component (expressed as multiples of p_{min}). All possible distributions of values for the nine different components (three per molecule) must be taken into account. The figure only shows one example of such a distribution.

component p_α, with $\alpha = x$, y, or z, gives the contribution $(p_\alpha)^2/2m$ to the energy since ε_{tr} for each molecule is given by Equation (E.1). The total energy is the sum of ε_{tr} for all molecules.

In a manner that corresponds to the arguments in Section 2.6, we will now determine how many different energy distributions there are among the molecules.[6] In doing so, we must take into account all possible distributions of energy between all $3N$ components for all molecules. It turns out to be easiest to first determine the number of possible energy distributions when the total energy is $\leq U$ and then from this result determine the number of distributions when the energy is U, which is the quantity we seek.

Let us first look at the case of a single momentum component p_α. The lowest energy is obtained when the component's absolute value is p_{min}, that is, the minimum energy is $\varepsilon_{min} = (p_{min})^2/2m$. If we allow energies between ε_{min} and U, the possible values of $|p_\alpha|$ are between p_{min} and p_{max}, which satisfies $(p_{max})^2/2m = U$, that is, $p_{max} = (2mU)^{1/2}$. All positive integer multiples of p_{min} are allowed, so the number of possible values of $|p_\alpha|$ is equal to $p_{max}/p_{min} = (U/\varepsilon_{min})^{1/2}$ (or more specifically, the integer closest below this number). Thus, the number of possible values is proportional to $\sqrt{U}$.

The derivation of the number of different energy distributions for N molecules, that is, for $3N$ momentum components, is given at the

---

[6]As before, we assume for simplicity that the molecules are distinguishable. This assumption does not affect, as we shall see, our final result in any significant way for those applications we shall consider.

end of this appendix (to be read by those who are particularly interested). It is shown that the number of possible energy distributions when the total energy is $\leq U$ is proportional to $(\sqrt{U})^{3N}$, provided that the energy is not very small. The number of distributions for $3N$ components is thus proportional to the corresponding number for a single component to the power $3N$. We have

$$\text{Number of distributions} = \text{const.} \times U^{3N/2},$$

where "const." is independent of U.

For a macroscopic system, N is a very large number, say, 10^{20} or greater. The number of energy distributions thus increases faster than $U^{10^{20}}$, which is incredibly fast. Even if U only increases by, say, one part per billion, the number of energy distributions with energies $\leq U$ increases by a factor that is greater than

$$(1.000000001)^{10^{20}} \approx 10^{4 \cdot 10^{10}}.$$

The number of energy distributions between $0.999999999\,U$ (that is, $U/1.000000001$) and U is thus more than $10^{4 \cdot 10^{10}}$ times greater than those between 0 and $0.999999999\,U$ (!). The distributions for energies $\leq 0.999999999\,U$ are thus negligibly few compared to those with energies in the immediate vicinity of U. Obviously, this is also the case when the system contains far fewer molecules than in 10^{20} and when the margin is much smaller than one part per billion. From this we can make the important conclusion that for macroscopic systems the number of energy distributions with energies $\leq U$ is in practice *equal to the number at the energy U.*[7]

In Section 2.6 we considered $\Omega(U)$, which is the number of possible energy distributions for total energy U of the system. From our previous argument we see that for a macroscopic system with N

---

[7]Since energies cannot be determined with infinite precision, it is actually not relevant to talk about the number of distributions with energy *exactly* equal to U. Therefore, the number of microstates with energies in the immediate vicinity of U, that we consider here, *is* the relevant quantity and constitutes $\Omega(U)$. One can show that $\ln \Omega$ is *extremely* insensitive to how wide the "immediate vicinity" is and one can actually even include *all* microstates with energy $\leq U$ in $\Omega(U)$ in calculations of $\ln \Omega$ for macroscopic systems at sufficiently high energies.

monatomic molecules, the following holds[8]

$$\Omega(U) = \mathcal{K}'U^{3N/2},$$ (E.2)

where $\mathcal{K}'$ is independent of U (the factor $\mathcal{K}'$ depends only on N, the volume V, and the mass m of a molecule and it is constant when only U varies). The entropy $S = k_B \ln \Omega$ for the monatomic gas is thus equal to[9]

$$S(U,V,N) = \frac{3}{2}Nk_B \ln U + S'(V,N),$$ (E.3)

where $S' = k_B \ln \mathcal{K}'$ is a contribution to the entropy that depends only on N and V (it includes the configurational entropy).

DERIVATION OF THE NUMBER OF ENERGY DISTRIBUTIONS*

We will here determine how many different distributions of energy exist for $3N$ momentum components. Let us initially allow the value of each component to vary between p_{min} and Jp_{min}, where J is an arbitrarily large positive integer. How many ways are there to realize a system where each component may independently adopt J different values?

As an example we take two components: the first component may assume the values $j_1 p_{min}$, $1 \le j_1 \le J$, and the second $j_2 p_{min}$, $1 \le j_2 \le J$, where j_1 and j_2 are integers (see Figure E.3a). When component 1 has the value p_{min} ($j_1 = 1$), component 2 has any of the values p_{min}, $2p_{min}$, ..., Jp_{min} ($1 \le j_2 \le J$). The same applies when component 1 has the value $2p_{min}$ ($j_1 = 2$). For each value of component 1, component 2 can thus assume J different values. In total there are thus $J \times J = J^2$ possibilities, which is the number of integer points within the square in Figure E.3a, that is, the area of the square.

---

[8]This result applies provided the energy is not very low, which we assume here. The value of $\mathcal{K}'$ depends on whether the molecules are treated as distinguishable or not, but otherwise the formula applies in both cases. In our applications of Equation (E.2) in Section 6.2 the value of $\mathcal{K}'$ does not matter.

[9]The complete expression for S of a monatomic ideal gas of indistinguishable particles is

$$S = k_B \ln \Omega(U,V,N) \approx k_B \ln \left[\left(\frac{4\pi m e}{3h^2} \right)^{3N/2} \left(\frac{U}{N} \right)^{3N/2} \frac{V^N}{N!} \right],$$

where e is the base of the natural logarithm, h is Planck's constant and the right-hand side is an excellent approximation for sufficiently large U and N (compare with footnote 9 in Section 5.2 and footnote 20 in Section 6.2).

This result can be easily generalized to more components. If one has three components, j_1 and j_2 can assume J^2 different values for each value that the third component adopts. There are thus $J^2 \times J = J^3$ possibilities. (This is the number of integer points within a cube with sides J and is equal to the volume of the cube.) For $3N$ components there are in an analogous manner J^{3N} possibilities.

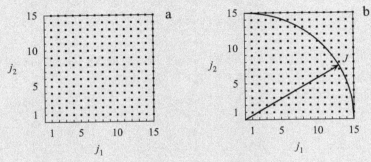

Figure E.3   An example of the various possibilities for two momentum components, which can have values $j_1 p_{min}$ and $j_2 p_{min}$ where j_1 and j_2 are positive integers. Each possibility for the two components is indicated by a dot. (a) The two components can each adopt J different values independently of each other, $1 \le j_1 \le J$ and $1 \le j_2 \le J$. In this example, $J = 15$. (b) If one has the condition that the energy at most can be $U = J^2 \varepsilon_{min}$, only the points within the radius J are included.

When we have a certain amount of energy to distribute, we are, however, not interested in all these possibilities. Let us therefore examine how many distributions of momentum there are when the total energy of the N molecules is lower than a certain value U. Each component p_α provides the contribution $(p_\alpha)^2/2m$ to energy. The smallest possible value of the energy contribution is $\varepsilon_{min} = (p_{min})^2/2m$. When the component has the value $|p_\alpha| = jp_{min}$, where j is a positive integer, its contribution to the energy is $(jp_{min})^2/2m = j^2 \varepsilon_{min}$.

We first examine the case of two components in Figure E.3. The sum of the energy contributions cannot exceed U, that is

$$[(j_1)^2 + (j_2)^2]\varepsilon_{min} \le U = J^2 \varepsilon_{min}, \tag{E.4}$$

where we have introduced

$$J = \left[\frac{U}{\varepsilon_{min}}\right]^{1/2} \tag{E.5}$$

(J is not necessarily an integer; as we shall see, this does not matter). We can express the condition (E.4) as $(j_1)^2 + (j_2)^2 \leq J^2$, which means that the possible points (j_1, j_2) lie within the radius J (see Figure E.3b), that is, within a quarter of a circle of radius J. The number of such points is proportional to the circle sector area (this is true at least as a very good approximation when J is a large number, i.e., when the energy is large enough). For two momentum components, the number of different distributions of energy $\leq U$ is thus proportional to J^2. The number of possibilities is growing with increasing J in the same way in both parts of Figure E.3 (both the area of the square and of the circle sector grow proportionally to J^2).

If we have three components, we saw earlier for the case corresponding to Figure E.3a that the number of possibilities is equal to J^3, which is the volume of a cube with the sides J. If we limit the energy $\leq U$ (which corresponds to the case of Figure E.3b) we have

$$[(j_1)^2 + (j_2)^2 + (j_3)^2]\varepsilon_{min} \leq U = J^2 \varepsilon_{min},$$

which implies that

$$(j_1)^2 + (j_2)^2 + (j_3)^2 \leq J^2,$$

where J is still given by Equation (E.5). Thereby, the possible points (j_1, j_2, j_3) lie within a sphere of radius J from the origin, as depicted in Figure E.4 (the points lie within the part that has positive coordinates, i.e., 1/8 sphere). The number of points grows with increasing J proportionally to the volume of the sphere,[10] which means the number grows proportionally to J^3.

We see that the number of possible energy distributions for both two and three components grows with increasing J as J^η, where η is the number of components, regardless of whether the energy is restricted to $\leq U = J^2 \varepsilon_{min}$ or the values of the components are restricted to $\leq J p_{min}$ independently of each other. This generally applies. For N molecules, we have $3N$ components and we saw earlier that for the case corresponding to Figure E.3a, the number of possibilities is

---

[10]More precisely, the number of possible distributions of energy $\leq U$ (the number of points) is, to an excellent approximation, equal to the volume of 1/8 sphere, that is, $\pi J^3/6$ when U and therefore J are large.

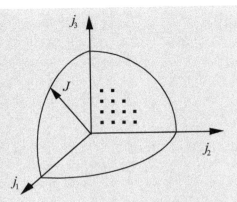

Figure E.4   The various possibilities for three momentum components are given by the values $j_\alpha p_{min}$, where j_α, $\alpha = 1, 2, 3$, are positive integers. The integer points (j_1, j_2, j_3) are spread out in space in the figure; only a few of them are drawn. When we have the condition that the energy must be $\leq U = J^2 \varepsilon_{min}$, only the points within the radius J are counted.

equal to J^{3N}. When we instead restrict the energy to $\leq U$, we have the condition

$$[(j_1)^2 + (j_2)^2 + (j_3)^2 + \ldots + (j_{3N})^2]\varepsilon_{min} \leq U, \tag{E.6}$$

which can be written

$$(j_1)^2 + (j_2)^2 + (j_3)^2 + \ldots + (j_{3N})^2 \leq J^2. \tag{E.7}$$

We cannot draw figures in $3N$ dimensions as illustrations, but also in this case the number of possibilities is proportional to J^{3N} (the "volume" of a so-called hypercube with sides J and the "volume" of a hypersphere of radius J grow both as J^{3N} when J increases).[11] Since J according to Equation (E.5) is proportional to $\sqrt{U}$, we can thus conclude that the number of energy distributions with energy $\leq U$ is growing like $U^{3N/2}$ when we increase U.

---

[11]In an analogy to the case of three dimensions (footnote 10), it follows that the number of possibilities (microstates) with energy $\leq U$ will be $1/2^{3N}$ times the "volume" of a hypersphere in $3N$ dimensions with radius J. (Incidentally, it can be mentioned that this together with $\varepsilon_{min} = (p_{min})^2/2m$ and $p_{min} = h/(2V^{1/3})$ [see footnote 2] can be used to obtain the expression for $\ln \Omega$ in footnote 9. It would, however, lead too far to give the details here.)

Symbols

Symbol	Explanation	SI unit
A	Helmholtz energy, $A = U - TS$	J
a	area	m^2
atm	pressure unit "atmosphere" ($1\,atm = 1.01325 \cdot 10^5\,Pa$)	$1\,Pa$ $= 9.869 \cdot 10^{-6}$ atm
bar	pressure unit ($1\,bar = 10^5\,Pa$)	$1\,Pa = 10^{-5}\,bar$
C	heat capacity	$J\,K^{-1}$
c	concentration (moles per unit volume), $c = n/V$	$mol\,m^{-3}$
c^0	concentration in the standard state; $c^0 = P^0/RT$ for ideal gas	$mol\,m^{-3}$
C_P	heat capacity at constant P	$J\,K^{-1}$
C_V	heat capacity at constant V	$J\,K^{-1}$
d	differential symbol; df is a small contribution to f	
dA, dS, dU, etc., see d above		
dq	small amount of heat (added to the system)	J
dw	small amount of work (done on the system)	J
F	force	$N = kg\,m\,s^{-2}$
G	Gibbs energy, $G = H - TS = U + PV - TS$	J

219

Symbol	Explanation	SI unit
G_m	Gibbs energy per mole of substance, $G_m = G/n$	$J\,mol^{-1}$
G^0	Gibbs energy for a substance in the standard state, whereby the pressure $= P^0$	J
G_m^0	Gibbs energy per mole of a substance in the standard state[1]	$J\,mol^{-1}$
H	enthalpy, $H = U + PV$	J
h	Planck's constant ($h = 6.6261 \cdot 10^{-34}\,J\,s$)	$J\,s$
H_m	enthalpy per mole of substance, $H_m = H/n$	$J\,mol^{-1}$
H^0	enthalpy for a substance in the standard state, whereby the pressure $= P^0$	J
H_m^0	enthalpy per mole of a substance in the standard state[1]	$J\,mol^{-1}$
K	equilibrium constant (the thermodynamic equilibrium constant)	dimensionless
k_B	Boltzmann's constant ($k_B = 1.3807 \cdot 10^{-23}\,J\,K^{-1}$)	$J\,K^{-1}$
K_c	equilibrium constant in concentration units, $K_c = K(c^0)^{\Delta N_r}$	$(mol\,m^{-3})^{\Delta N_r}$
K_P	equilibrium constant in pressure units, $K_P = K(P^0)^{\Delta N_r}$	$(Pa)^{\Delta N_r}$
M	concentration unit "molar" $= mol\,dm^{-3}$ ($1\,M = 10^3\,mol\,m^{-3}$)	$1\,mol\,m^{-3} = 10^{-3}\,M$
N	number of particles or molecules	unitless
n	number of moles of a substance, $n = N/N_{Av}$	mol
N_{Av}	Avogadro's constant ($N_{Av} = 6.02214 \cdot 10^{23}\,mol^{-1}$)	mol^{-1}

Symbol	Explanation	SI unit
N_i	number of molecules of species i	unitless
n_i	number of moles of species i, $n_i = N_i/N_{Av}$	mol
P	pressure = force per unit area	Pa
$\mathbf{p}$	momentum vector, $\mathbf{p} = m\mathbf{v}$	$\mathrm{kg\,m\,s^{-1}} = \mathrm{N\,s}$
p	probability	unitless
Pa	pressure unit "Pascal" = $\mathrm{N\,m^{-2}}$	Pa
P^{ext}	external pressure	Pa
P_i	partial pressure of species i, $P_i = x_i P$	Pa
P^0	pressure at the standard state (usually $P^0 = 1\,\mathrm{bar} = 10^5\,\mathrm{Pa}$)	Pa
Q	reaction quotient	unitless
q	heat (added to the system)	J
Q^{eq}	the value of Q at equilibrium	unitless
q_{surr}	heat added to the surroundings	J
R	the universal gas constant ($R = k_B N_{Av} = 8.3145\,\mathrm{J\,K^{-1}mol^{-1}}$)	$\mathrm{J\,K^{-1}mol^{-1}}$
S	entropy	$\mathrm{J\,K^{-1}}$
s	distance	m
S_{conf}	entropy from number of particle configurations	$\mathrm{J\,K^{-1}}$
S_{ener}	ΔS_{ener} and dS_{ener} are entropy changes due to spreading of energy	$\mathrm{J\,K^{-1}}$
S_{irrev}	$\Delta S_{\mathrm{irrev}}$ and dS_{irrev} are entropy changes due to irreversible processes	$\mathrm{J\,K^{-1}}$
S_m	entropy per mole of substance, $S_m = S/n$	$\mathrm{J\,K^{-1}mol^{-1}}$
S_{surr}	entropy of the surroundings	$\mathrm{J\,K^{-1}}$

Symbol	Explanation	SI unit
S_{tot}	total entropy of the system and the surroundings, $S_{tot} = S + S_{surr}$	$J K^{-1}$
S^0	entropy for a substance in the standard state, whereby the pressure = P^0	$J K^{-1}$
S_m^0	entropy per mole of a substance in the standard state; the standard entropy	$J K^{-1} mol^{-1}$
T	absolute temperature	K
T_b	boiling point temperature	K
T_m	melting point temperature	K
U	internal energy	J
V	volume	m^3
v	speed (length of velocity vector), $v = |\mathbf{v}|$	$m s^{-1}$
$\mathbf{v}$	velocity vector, $\mathbf{v} = (v_x, v_y, v_z)$	$m s^{-1}$
w	work (done on the system)	J
w_{surr}	work done on the surroundings	J
x_i	mole fraction of substance i; for a mixture of substances 1, 2, and 3: $x_i = N_i/(N_1 + N_2 + N_3) = n_i/(n_1 + n_2 + n_3)$	unitless
Δ	change of some quantity; Δf is the change of f	
$\Delta A, \Delta S, \Delta U$, etc., see Δ above		
ΔN_r	change in number of molecules during a reaction as given by the stoichiometric coefficients (number of product molecules – reactant molecules)	unitless
$\Delta_f G_m^0$	Gibbs energy of formation per mole of a substance[1]	$J mol^{-1}$

Symbol	Explanation	SI unit
$\Delta_f H_m^0$	enthalpy of formation per mole of a substance[1]	$J\,mol^{-1}$
Δ_{fus}	difference between the liquid and the solid states for some quantity	
Δ_r	change of some quantity during a reaction according to the stoichiometry of the reaction formula (in moles)	
$\Delta_r G$	ΔG during a reaction when reactants and products are present in such a large amount in the reaction mixture that the changes in their concentrations are negligible	$J\,mol^{-1}$
$\Delta_r G^0$	ΔG during a reaction where reactants and products are pure substances in their standard states (see Figure 5.6)	$J\,mol^{-1}$
$\Delta_r H^0$	see $\Delta_r G^0$ (but with H instead of G)	$J\,mol^{-1}$
$\Delta_r S^0$	see $\Delta_r G^0$ (but with S instead of G)	$J\,K^{-1}mol^{-1}$
Δ_{vap}	difference between the vapor and the liquid states for some quantity	
ε_{tr}	kinetic (translational) energy for a molecule, $\varepsilon_{tr} = mv^2/2$	J
$\bar{\varepsilon}_{tr}$	average of ε_{tr} over all molecules	J
μ	chemical potential ($\mu = G_m$ for pure substance)	$J\,mol^{-1}$
ν	cell volume	m^3
ρ	particle (molecular) density, $\rho = N/V$	m^{-3}
ρ^0	particle (molecular) density in the standard state; $\rho^0 = P^0/k_B T$ for ideal gas	m^{-3}
ρ_i	particle (molecular) density for species i, $\rho_i = N_i/V$	m^{-3}

Symbol	Explanation	SI unit
Ω	number of microstates	unitless

---

[1]The molar Gibbs energy G_m^0 and enthalpy H_m^0 for a pure substance in the standard state are equal to the Gibbs energy and enthalpy of formation, $\Delta_f G_m^0$ and $\Delta_f H_m^0$, respectively, provided the zero levels of H and G are set according to the convention "for a pure element in its most stable form at the temperature in question $H = 0$ and $G = 0$ in the standard state." This is the convention used in the book; see the discussion in Section 5.1.

Index

Note: Numbers in *italics* indicate pages with relevant figures and numbers in **bold** indicate the main pages for the topic.